Solutions Manual to Accompany Introduction to Quantitative Methods in Business

Solutions Manual to Accompany Introduction to Quantitative Methods in Business:
With Applications Using Microsoft® Office Excel®

Bharat Kolluri
Michael J. Panik
Rao N. Singamsetti

WILEY

Published by John Wiley & Sons, Inc., Hoboken, New Jersey
Published simultaneously in Canada

For general information on our other products and services or for technical support, please contact our Customer Care Department within the United States at (800) 762-2974, outside the United States at (317) 572-3993 or fax (317) 572-4002.

Wiley also publishes its books in a variety of electronic formats. Some content that appears in print may not be available in electronic formats. For more information about Wiley products, visit our web site at www.wiley.com.

Library of Congress Cataloging-in-Publication Data

Names: Kolluri, Bharat, author. | Panik, Michael J., author. | Singamsetti, Rao, author.
Title: Introduction to quantitative methods in business : with applications using Microsoft Office Excel / Bharat Kolluri, Michael J. Panik, Rao Singamsetti.
Description: Hoboken, New Jersey : John Wiley & Sons, Inc., [2017] | Includes index.
Identifiers: LCCN 2016017312 | ISBN 9781119220978 (cloth) | ISBN 9781119221029 (solutions manual) | ISBN 9781119220992 (epub)
Subjects: LCSH: Business mathematics. | Management–Mathematical models. | Microsoft Excel (Computer file)
Classification: LCC HF5691 .K71235 2017 | DDC 650.0285/54–dc23 LC record available at https://lccn.loc.gov/2016017312

10 9 8 7 6 5 4 3 2 1

Table of Contents

Note to the reader: In this Manual the solutions to the odd-numbered exercises for each chapter are preceded by a summary of the requisite material presented in the main text. Hence the numbering of the sections offered herein mirrors those used to designate the various portions of the textbook.

Chapter 1

The Mathematical Toolbox: A Summary

1.2 LINEAR FUNCTIONS

An expression such as $Y = b_0 + b_1 X$ represents a *linear equation (function)*, where b_0 is the Y-intercept (it gives the value of Y when $X = 0$) and $b_1 = \Delta Y / \Delta X$ is the *slope* (often referred to as *rise/run*). Here Y is the *dependent variable* and X is the *independent variable*. Note that both b_0 and b_1 are constants.

1.3.1 Solving Two Simultaneous Linear Equations

At times you will need to obtain a solution to a set of simultaneous linear equations, that is, to a set of equations of the general form:

$$aX + bY = e, \tag{1.1}$$

$$cX + dY = f. \tag{1.2}$$

A system such as this is said to be *consistent* if it has at least one solution. Moreover, if $ad - cb \neq 0$, then this equation system is consistent. For instance, the equation system

$$X - Y = 6 \tag{1.3}$$

$$3X - 2Y = 4 \tag{1.4}$$

is consistent since (1) $(-2) - (3) (-1) = 1 \neq 0$. In fact, to obtain the (unique) solution, we can multiply Equation (1.3) by -3 so as to obtain $-3X + 3Y = -18$, and then add this multiple to Equation (1.4) to get $Y = -14$. If we then substitute $Y = -14$ into Equation (1.3), we obtain $X = -8$. How do we know that we have generated the correct solution to this equation system?

Solutions Manual to Accompany Introduction to Quantitative Methods in Business: With Applications Using Microsoft® Office Excel®, First Edition. Bharat Kolluri, Michael J. Panik, and Rao N. Singamsetti.
© 2017 John Wiley & Sons, Inc. Published 2017 by John Wiley & Sons, Inc.

Answer: Substitute $X = -8$ and $Y = -14$ back into, say, Equation (1.4) and show that equality holds.

It is easily demonstrated that the equation system

$$4X_1 + 2X_2 = 3,$$
$$16X_1 + 8X_2 = 12,$$

is *inconsistent* or *dependent* in that it has no solution. Here $(4)(8) - (16)(2) = 0$. Clearly, these two equations represent parallel lines—they do not intersect.

1.4 SUMMATION NOTATION

The operation of addition of a set of n values is readily carried out by using the "sigma" notation. In this regard, the left-hand side of the expression

$$\sum_{i=1}^{n} X_i = X_1 + X_2 + \cdots + X_n$$

reads: "the sum of all values X_i as i goes from 1 to n." The right-hand side shows that the operation of addition has been executed. Some useful summation rules are as follows:

Rule 1: $\sum_{i=1}^{n} (X_i \pm Y_i) = \sum_{i=1}^{n} X_i \pm \sum_{i=1}^{n} Y_i$.

Rule 2: $\sum_{i=1}^{n} cX_i = c \sum_{i=1}^{n} X_i$, where c is a constant.

Rule 3: $\sum_{i=1}^{n} c = nc$, where c is a constant.

Note also that

$$\sum_{i=1}^{n} X_i^2 \neq \left(\sum_{i=1}^{n} X_i \right)^2,$$

$$\sum_{i=1}^{n} X_i Y_i \neq \left(\sum_{i=1}^{n} X_i \right) \left(\sum_{i=1}^{n} Y_i \right),$$

if $\overline{X} = \sum_{i=1}^{n} X_i / n$ is the sample mean, then

$$\sum_{i=1}^{n} (X_i - \overline{X}) = 0.$$

The *Pearson sample correlation coefficient* can be written as either

$$r = \frac{\sum_{i=1}^{n} (X_i - \overline{X})(Y_i - \overline{Y})}{\sqrt{\sum_{i=1}^{n} (X_i - \overline{X})^2 \sum_{i=1}^{n} (Y_i - \overline{Y})^2}} \quad \text{(long formula)},$$

where $\overline{X} = \sum_{i=1}^{n} X_i / n$ and $\overline{Y} = \sum_{i=1}^{n} Y_i / n$ are the sample means of X and Y, respectively; or as

$$r = \frac{\sum X_i Y_i - \frac{\sum X_i \sum Y_i}{n}}{\left[\left(\sum X_i^2 - \left[(\sum X_i)^2 / n \right] \right) \left(\sum Y_i^2 - \left[(\sum Y_i)^2 / n \right] \right) \right]^{1/2}} \quad \text{(short formula)}.$$

Given a collection of data points (X_i, Y_i), $i = 1, \ldots, n$, we can fit a linear equation of the form $Y = a + bX$ through them, where a is the Y-intercept and b is the slope. Here,

$$b = \frac{\sum (X_i - \overline{X})(Y_i - \overline{Y})}{\sum (X_i - \overline{X})^2} \quad \text{(long formula)}$$

or

$$b = \frac{\sum X_i Y_i - \frac{\sum X_i \sum Y_i}{n}}{\left[\sum X_i^2 - \frac{(\sum X_i)^2}{n} \right]}$$

and $a = \overline{Y} - b\overline{X}$.

1.5   SETS

We know that a *set* is a collection or grouping of items (called *elements*) without regard to structure or order and that a set is usually defined by listing its elements. A set containing no elements is called the *null set* or the *empty set* and is denoted as ϕ. A set containing all elements under a particular discussion or in a given context is termed the *universal set* and is denoted as U. Given a set A, its complement, denoted $\overline{A}$, is the set containing all the elements within U that lie outside of A.

The *intersection* of two sets A and B is the set of elements common to both A and B. It is denoted as $A \cap B$. The *union* of two sets A and B contains the elements in A, or in B, or in both A and B. It is denoted as $A \cup B$. A moments reflection reveals the following:

$$\overline{\phi} = U$$
$$\overline{U} = \phi$$
$$A \cup \phi = A$$
$$A \cap \phi = \phi$$
$$A \cup U = U$$
$$A \cap U = A$$
$$A \cup \overline{A} = U$$
$$A \cap \overline{A} = \phi$$

1.6   FUNCTIONS AND GRAPHS

A *function f* is a rule or law of correspondence that associates with each value of a variable x a unique value of a variable y. Here y is termed the *image* of x under rule f. This "rule" is written as $y = f(x)$ and is read: "y is a function of x." Think

x	y = f(x) = x²
0	0
1	1
2	4
3	9
4	16

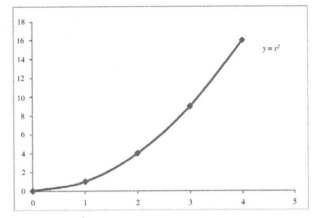

Figure 1.1   The graph of the function $y = f(x) = x^2$.

of a function as a recipe for getting unique y values from x values. Here x is termed the *independent variable* and y is called the *dependent variable*. The set of all admissible x values is called the *domain* of the function; the collection of y values, which are the image of at least one x, is termed the *range* of the function. For instance, a function may appear as $y = f(x) = x^2 + 2$. What is the rule that is operative here? The rule is: Select a value of x, multiply it by itself, and then add 2 to the result to get y. So, if $x = 2$, the image of $x = 2$ via rule f is

$$y = (2)^2 + 2 = 6.$$

It is important to note that for each x value there is one and only one resulting value of y. However, a y value may be the image of more than one x value. However, if $x_1 \neq x_2$ implies $f(x_1) \neq f(x_2)$, then f is said to be a *one-to-one* function.

A useful device for illustrating a function is its *graph*. Here y is plotted on the vertical axis and x is plotted on the horizontal axis. Then, via the rule f, a set of x values are chosen and their corresponding y values are determined. These ordered (x, y) combinations are then connected to form the graph of the function in the x–y-plane (see Figure 1.1).

Given a particular graph, how can we determine whether or not it represents a function?

Answer: We can employ the *vertical line test*: For a given value of x, find y by drawing a vertical line at x. If the line cuts the graph at only a single value of y, then the graph indeed represents a function.

1.7 WORKING WITH FUNCTIONS

At times it is important to be able to graph a linear function in a quick and efficient fashion. To this end, let us consider the following approach:

> *Intercept method:* Since two points are enough to draw a unique straight line, find both the *horizontal intercept* (set $y = 0$ and solve for the

x-intercept) and the *vertical intercept* (set $x = 0$ and solve for the y-intercept) and connect the resulting two points, which are called *basic points*. For instance, suppose $y = 2x - 6$. Then,

a. if $y = 0$, then $2x - 6 = 0$ or $x = 3$; and
b. if $x = 0$, then $y = -6$.

Hence, the two basic points are (3, 0) and (0, −6). These points are then connected to obtain the desired graph.

An alternative method for graphing a function (which is useful for both linear and nonlinear equations) is the point method:

Point method: Given an admissible set of x values, we use the rule f to get the corresponding y values and then we connect each of the resulting ordered pairs (x, y) (see Figure 1.1).

1.8 DIFFERENTIATION AND INTEGRATION

We may think of a function as being *continuous* if its graph is without breaks. Additionally, the *derivative of a function $y = f(x)$ with respect to x* is defined as

$$\lim_{\Delta x \to 0} \frac{\Delta y}{\Delta x} = \frac{dy}{dx} = f'(x).$$

Here dy/dx is the instantaneous rate of change in y per unit change in x as Δx gets smaller and smaller. Note that if f has a derivative at a point $x = a$, then it is also continuous at $x = a$. However, if f is continuous at a point $x = a$, it does not follow that f has a derivative there.

The derivative $f'(x) = dy/dx$ is also interpreted as the slope of the function $y = f(x)$ at a given point, that is, for $x = a$, $f'(a)$ is the slope of f at this x value. Additionally, dy/dx depicts the *marginal change in y with respect to x*.

Rules of Differentiation:

1. *Power rule:* If $y = f(x) = x^n$, then $dy/dx = nx^{n-1}$. (If $y = x^4$, $dy/dx = 4x^3$.)

2. The derivative of a constant is zero.

3. *Coefficient rule:* If $y = f(x) = kg(x)$, then $dy/dx = k(dg/dx)$. (If $y = 6x^2$, then $dy/dx = 12x$. Here, $k = 6$ and $g(x) = x^2$.)

4. *Sum (difference) rule:* If $y = f(x) = g(x) \pm h(x)$, then $dy/dx = dg/dx \pm dh/dx$. (If $y = 2x^2 - 6x^3$, then $dy/dx = 4x - 18x^2$.)

5. *Product rule:* If $y = f(x) = g(x) \times h(x)$, then $dy/dx = (dg/dx) \times h(x) + (dh/dx) \times g(x)$. (If $y = (3x - 2)(2x^2 + 8)$, then $dy/dx = (3)(2x^2 + 8) + (3x - 2)(4x)$.)

6. *Quotient rule:* If $y = f(x) = g(x)/h(x)$, $h(x) \neq 0$, then

$$\frac{dy}{dx} = \frac{h(x) \times (dg/dx) - g(x) \times (dh/dx)}{(h(x))^2}.$$

(If $y = \dfrac{10x - x^2}{2x^3 + 6}$, then $\dfrac{dy}{dx} = \dfrac{(2x^3 + 6)(10 - 2x) - (10x - x^2)(6x^2)}{(2x^3 + 6)^2}$.)

7. Derivatives of exponential and logarithmic functions:

a. If $y = e^x$, $dy/dx = e^x$.

b. If $y = e^{kx}$, $dy/dx = ke^{kx}$, where k is a constant.

c. If $y = \ln x$, $dy/dx = \frac{1}{x}$.

Higher Order Derivatives:

1. Given that the *first derivative of f with respect to* x appears as $f'(x) = dy/dx$, the *second derivative of f with respect to* x is written as $f''(x) = d^2y/dx^2$ and is calculated as the derivative of the first derivative.

   (If $y = 3x^3 + 6x^2 - 2$, $f'(x) = 9x^2 + 12x$, and $f''(x) = 18x + 12$.)

   Geometrically, the second derivative of f is the rate of change of the slope of f at a specific point on the graph of f. For instance, at a point $x = a$, if $f''(a) < 0$, then f is said to be *concave downward at* $x = a$; and if $f''(a) > 0$, then f is termed *concave upward of* $x = a$.

2. The *third derivative of f with respect to* x is the derivative of the second derivative. (Given $f''(x) = 18x + 12$ above, $f''(x) = 18$.)

3. In general, the *nth order derivative of f with respect to* x is the derivative of the $(n - 1)$th order derivative, for example, the sixth derivative of f is the derivative of the fifth derivative of f.

There are two distinct ways to view the concept of an integral. On the one hand, we have the *definite integral*—an integral that is the limit of an addition process and can be thought of as an area under a given curve. On the other hand, we have the *indefinite integral*—an integral determined as the result of reversing the process of differentiation.

1. *Definite integral:* Given the derivative $dy/dx = f(x)$, $a < x < b$, we need to find $y = F(x)$. In this regard, the function $y = F(x)$ is a solution to $dy/dx = f(x)$ if $F(x)$ is differentiable over $a < x < b$ and $dF(x)/dx = f(x)$. Hence, $F(x)$ is the *integral of f(x) with respect to x*. So, if $y = F(x)$ is a particular solution of $dy/dx = f(x)$, then "all" solutions can be written as $y = F(x) + c$, where c is an arbitrary constant. This is indicated by the expression $\int f(x)\,dx = F(x) + c$.

2. *Definite integral:* Suppose we wish to determine the area under the continuous curve $y = f(x)$ and above the x-axis for $a \leq x \leq b$. This area, called the *definite integral of f (x)*, is the limiting sum of a set of

rectangles as the number of rectangles increases without bound and is denoted as

$$\int_a^b f(x)dx.$$

This integral is determined via the *fundamental theorem of integral calculus*: If $f(x)$ is continuous on $a \le x \le b$ and F is any differentiable function such that $F'(x) = f(x)$, $a \le x \le b$, then

$$\int_a^b f(x)dx = F(x)]_a^b = F(b) - F(a).$$

Rules of Integration:

1. *Power rule:*

$$\int x^n dx = \frac{x^{n+1}}{n+1} + c, \quad n \ne -1$$

if $n = -1$,

$$\int x^{-1} dx = \ln x + c.$$

(Here, $\int x^3 dx = \frac{1}{4}x^4 + c$.)

2. The integral of zero is a constant or $\int 0 dx = c$.

3. The integral of a constant times a function is the constant times the integral of the function or $\int kf(x)dx = k\int f(x)dx$, where k is a constant.
(Here, $\int 3x^5 dx = 3\int x^5 dx = 3\left(\frac{1}{6}x^6\right) + c$.)

4. Integral of a sum or difference rule:

$$\int (f(x) \pm g(x))dx = \int f(x)dx \pm \int g(x)dx.$$

Here,

$$\int (8x^2 + 3x - 6)dx = 8\int x^2 dx + 3\int xdx$$

$$-6\int dx = 8\left(\frac{1}{3}x^3\right) + 3\left(\frac{1}{2}x^2\right) - 6x + c.$$

5. *Integration by parts:* $\int u dv = uv - \int v du.$

Find $\int x(x+3)^3 dx$. Set $u = x$ and $dv = (x+3)^3 dx$. Then, $du = 1$ and $v = \int (x+3)^3 dx = \frac{1}{4}(x+3)^4$. Hence, from $\int u\,dv = uv - \int v\,du$,

$$
\begin{aligned}
\int x(x+3)^3 dx &= x\left(\frac{1}{4}(x+3)^4\right) - \frac{1}{4}\int (x+3)^4 dx \\
&= \frac{1}{4}x(x+3)^4 - \frac{1}{4}\left(\frac{1}{5}\right)(x+3)^5 + c.
\end{aligned}
$$

6. Integral of an exponential or logarithmic function:

    a. $\int e^x dx = e^x + c.$

    b. $\int e^{kx} dx = \frac{1}{k}e^{kx} + c$, where k is a constant.

    c. $\int \ln x\, dx = x(\ln x - 1) + c.$

SOLUTIONS TO ODD-NUMBERED EXERCISES

Solve the following:

1. $\frac{1}{3}X + 7 = 51$

3. $2.3X - 8.9 = -4.3$

SOLUTION:

  1. $\frac{1}{3}X = 51 - 7$

    $\frac{1}{3}X = 44$

    $X = 3 \times 44 = 132$

SOLUTION:

  3. $2.3X = -4.3 + 8.9$

    $2.3X = 4.6$

    $X = \dfrac{4.6}{2.3} = 2$

Solve the following sets of simultaneous equations:

  5. $2X + 3Y = 18$
       $X + 3Y = 15$

  7. $2X_1 - X_2 = 3$
       $7X_1 + 2X_2 = 27$

SOLUTION:

  5. Designate the first equation as (1) and the second equation as (2)

$$2X + 3Y = 18 \tag{1}$$

$$X + 3Y = 15 \tag{2}$$

Subtracting (2) from (1), we get

$$X = 3$$

Substitute the value of X in any of the above equations, for example,

$$X + 3Y = 15$$
$$3Y = 15 - 3 = 12$$
$$Y = \frac{12}{3} = 4$$

Check of (1):

$$2(3) + 3(4) = 6 + 12 = 18$$

Check of (2):

$$3 + 3(4) = 3 + 12 = 15$$

SOLUTION:

7. Designate the first equation as (1) and the second equation as (2):

$$2X_1 - X_2 = 3 \tag{1}$$
$$7X_1 + 2X_2 = 27 \tag{2}$$

Multiply (1) by 2 and call the resulting equation (3). Then, adding equations (3) and (2), we get

$$7X_1 + 2X_2 = 27 \tag{2}$$
$$\underline{4X_1 - 2X_2 = 6} \tag{3}$$
$$11X_1 = 33$$

$$X_1 = \frac{33}{11} = 3$$

Now substitute the value of X_1 into equation (1) to get

$$2(3) - X_2 = 3$$
$$-X_2 = 3 - 6 = -3, \ X_2 = 3$$

Check of (1):

$$2(3) - 3 = 6 - 3 = 3$$

Check of (2):

$$7(3) + 2(3) = 21 + 6 = 27$$

Suppose that there are six observations for the variables X and Y as given below:

$$X_1 = 2, \ X_2 = 1, \ X_3 = 3, \quad X_4 = -5, \ X_5 = 1, \ X_6 = -2$$
$$Y_1 = 4, \ Y_2 = 0, \ Y_3 = -1, \ Y_4 = 2, \quad Y_5 = 7, \ Y_6 = -3$$

For Exercises 9, 11, 13, 15, 17, and 19, compute the following:

9. $\sum_{i=1}^{6} X_i$

11. $\sum_{i=1}^{6} X_i^2$

13. $\sum_{i=1}^{6} X_i Y_i$

15. $\sum_{i=1}^{6} (X_i - Y_i)$

17. $\sum_{i=1}^{6} cX_i,$ where $c = -1$

19. $\sum_{i=1}^{6} (X_i - Y_i)^2$

$$= \sum_{i=1}^{6} X_i^2 - \sum_{i=1}^{6} 2X_i Y_i + \sum_{i=1}^{6} Y_i^2$$

SOLUTION:

9. $\sum_{i=1}^{6} X_i = X_1 + X_2 + X_3 + X_4 + X_5 + X_6$

$$= 2 + 1 + 3 + (-5) + 1 + (-2) = 7 - 7 = 0$$

SOLUTION:

11. $\sum_{i=1}^{6} X_i^2 = X_1^2 + X_2^2 + X_3^2 + X_4^2 + X_5^2 + X_6^2$

$$= (2)^2 + (1)^2 + (3)^2 + (-5)^2 + (1)^2 + (-2)^2$$

$$= 4 + 1 + 9 + 25 + 1 + 4 = 44$$

SOLUTION:

13. $\sum_{i=1}^{6} X_i Y_i = X_1 Y_1 + X_2 Y_2 + X_3 Y_3 + X_4 Y_4 + X_5 Y_5 + X_6 Y_6$

$$= (2)(4) + (1)(0) + (3)(-1) + (-5)(2) + (1)(7) + (-2)(-3)$$

$$= 8 + 0 - 3 - 10 + 7 + 6 = 8$$

SOLUTION:

15. $= (X_1 - Y_1) + (X_2 - Y_2) + (X_3 - Y_3) + (X_4 - Y_4) + (X_5 - Y_5) + (X_6 - Y_6)$

$$= (2 - 4) + (1 - 0) + (3 + 1) + (-5 - 2) + (1 - 7) + (-2 + 3)$$

$$= -2 + 1 + 4 - 7 - 6 + 1 = -9$$

SOLUTION:

17. $c\sum_{i=1}^{6} X_i = -1 \times 0 = 0$ since $c = -1$ and $\sum_{i=1}^{6} X_i = 0$

SOLUTION:

19. The given expression can be written as

$$\sum_{i=1}^{6} X_i^2 - 2\sum_{i=1}^{6} X_i Y_i + \sum_{i=1}^{6} Y_i^2$$

We already know that

$$\sum_{i=1}^{6} X_i^2 = 44$$

$$\sum_{i=1}^{6} X_i Y_i = 8$$

$$\sum_{i=1}^{6} Y_i^2 = 79$$

Substituting these values in the above equation yields

$$44 - 2 \times 8 + 79 = 44 - 16 + 79 = 107$$

21. Write set notation for the set of odd whole numbers greater than 0.

SOLUTION:

21. $\{1,3,5,7,9, \ldots\}$

Determine whether true or false in Exercises 23 and 25.

23. $2 \in \{-3, 2, 4, 5, 7\}$

25. $\{0, 1, 7\} \subset \{0, 1, 8, 13\}$

SOLUTION:

23. True

SOLUTION:

25. False, because set $\{0,1,7\}$ is not a subset of the $\{0, 1, 8, 13\}$.

27. Find $\{3,5,9\} \cup \{1,6,7\}$

SOLUTION:

27. $\{1, 3, 5, 6, 7, 9\}$

29. Using the information
$U = \{3 \text{ red marbles}, 7 \text{ blue marbles}, 8 \text{ white marbles}\}$ and $A = \{8 \text{ white marbles}\}$, find $A \cup \overline{A}$, $A \cap \overline{A}$, $\overline{U}$, and $\overline{\phi}$.

SOLUTION:

29. $A \cup \overline{A} = \{8 \text{ white marbles}, 3 \text{ red marbles}, 7 \text{ blue marbles}\}$

$$A \cap \overline{A} = \phi$$

$$\overline{U} = \phi$$

$$\overline{\phi} = U$$

Graph the following functions:

31. $2x + y - 10 = 0$

33. $y - 3\sqrt{x} = 5$

SOLUTION:

31. We can write this equation in standard form as $y = 10 - 2x$; prepare a table of some arbitrary x values and the corresponding y values and plot the points on a graph and connect them to get a straight line as shown below:

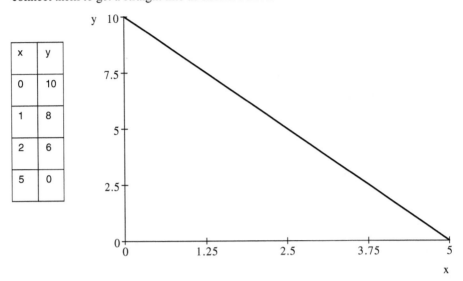

x	y
0	10
1	8
2	6
5	0

SOLUTION:

33.

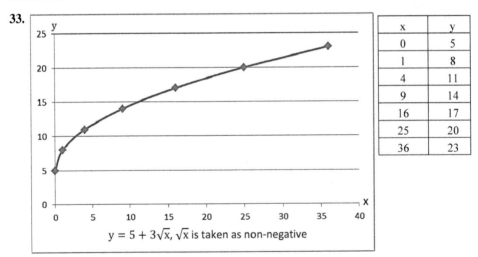

x	y
0	5
1	8
4	11
9	14
16	17
25	20
36	23

$y = 5 + 3\sqrt{x}$, $\sqrt{x}$ is taken as non-negative

Find dy/dx for the following functions given in Exercises 35, 37, 39, 41, and 43.

35. $y = 1/x^3$, $x \neq 0$. Evaluate dy/dx at $x = 1$.

37. $y = (1/2)$

39. $y = (1/5)x^5$

41. $y = -0.5x + 3$

43. $y = 5x^3 + 4x^2 + 3x + 2$

SOLUTION:

35. $\dfrac{dy}{dx} = \dfrac{d(x^{-3})}{dx} = -3x^{-3-1} = -3x^{-4} = -3\left(\dfrac{1}{X^4}\right)$. At x = 1, $\dfrac{dy}{dx} = -3$

SOLUTION:

37. $\dfrac{dy}{dx} = 0$

SOLUTION:

39. $\dfrac{dy}{dx} = \dfrac{1}{5}\dfrac{d(x^5)}{dx} = \dfrac{1}{5} \times 5x^4 = x^4$

SOLUTION:

41. $\dfrac{dy}{dx} = -0.5\dfrac{d(x)}{dx} + 0 = -0.5$

SOLUTION:

43. $\dfrac{dy}{dx} = 5\dfrac{d(x^3)}{dx} + 4\dfrac{d(x^2)}{dx} + 3\dfrac{d(x)}{dx} + 0$

$\qquad = 5 \times 3x^2 + 4 \times 2x + 3 = 15x^2 + 8x + 3$

45. **(a)** If $f(x) = 3x - 6$, $g(x) = 4x^2 + x$, and $y = f(x)/g(x)$, find $\dfrac{dy}{dx}$.

(b) If $f(x) = 5x^3 - 10$, $g(x) = -x^2 + x$, and $y = \dfrac{f(x)}{g(x)}$, find $\dfrac{dy}{dx}$.

SOLUTION:

45. **(a)** $\dfrac{dy}{dx} = \dfrac{g(x)f'(x) - f(x)g'(x)}{(g(x))^2} = \dfrac{(4x^2 + x)(3) - (3x - 6)(8x + 1)}{(4x^2 + x)^2}$

$\qquad = \dfrac{12x^2 + 24x + 6}{16x^4 + 8x^3 + x^2}$

(b) $\dfrac{dy}{dx} = \dfrac{g(x)f'(x) - f(x)g'(x)}{(g(x))^2} = \dfrac{(-x^2 + x)(15x^2) - (5x^3 - 10)(-2x + 1)}{(-x^2 + x)^2}$

$\qquad = \dfrac{-5x^4 + 10x^3 - 20x + 10}{-x^4 - 2x^3 + x^2}$

47. $\displaystyle\int \dfrac{1}{x^2}\, dx$

SOLUTION:

47. $\displaystyle\int x^{-2}dx = \dfrac{1}{(-2 + 1)}x^{-2+1} + c$, where c is a constant

$\qquad = -1x^{-1} + c = \dfrac{-1}{x} + c$

49. $\int \left(3x^2 - 4x + 8\right)dx$

SOLUTION:

49. $\int 3x^2 dx - \int 4x dx + \int 8 dx = \dfrac{3x^{2+1}}{(2+1)} + c_1 - \dfrac{4x^{1+1}}{(1+1)} + c_2 + 8x + c_3$

$\qquad = x^3 - 2x^2 + 8x + c, \quad \text{where } c = c_1 + c_2 + c_3 \ (c_1, c_2, \text{ and } c_3 \text{ are constants})$

51. $\displaystyle\int_a^b \dfrac{1}{(b-a)}dx$

SOLUTION:

51. $\left[\dfrac{1}{(b-a)}x\right]_a^b = \dfrac{b}{b-a} - \dfrac{a}{b-a} = \dfrac{b-a}{b-a} = 1$

53. Given

$\qquad X_1 = -2, \ X_2 = -3, \ X_3 = 0, \ X_4 = 5, \ X_5 = 9$

$\qquad Y_1 = 5, \quad Y_2 = 9, \quad Y_3 = 6, \ Y_4 = 8, \ Y_5 = 7$

(a) Evaluate the Pearson correlation coefficient

$$r = \sum_{i=1}^{5} \dfrac{\left(X_i - \overline{X}\right)\left(Y_i - \overline{Y}\right)}{\sqrt{\sum_{i=1}^{5}\left(X_i - \overline{X}\right)^2 \sum_{i=1}^{5}\left(Y_i - \overline{Y}\right)^2}}, \quad (-1 \le r \le 1)$$

where $\overline{X} = \sum X_i / 5$ and $\overline{Y} = \sum Y_i / 5$. How is r interpreted?

(b) Also, re-evaluate r using the following "short formula":

$$r = \dfrac{\sum X_i Y_i - \dfrac{\sum X_i \sum Y_i}{n}}{\left[\left(\sum X_i^2 - \dfrac{\left(\sum X_i\right)^2}{n}\right)\left(\sum Y_i^2 - \dfrac{\left(\sum Y_i\right)^2}{n}\right)\right]^{\frac{1}{2}}}$$

(*Hint:* To facilitate your calculations, construct a table.)

SOLUTION:

53. (a)

Serial No.	X_i	Y_i	$\left(X_i - \overline{X}\right)$	$\left(Y_i - \overline{Y}\right)$	$\left(X_i - \overline{X}\right) \times \left(Y_i - \overline{Y}\right)$	$\left(X_i - \overline{X}\right)^2$	$\left(Y_i - \overline{Y}\right)^2$
1	−2	5	−3.8	−2	7.6	14.44	4
2	−3	9	−4.8	2	−9.6	23.04	4
3	0	6	−1.8	−1	1.8	3.24	1
4	5	8	3.2	1	3.2	10.24	1
5	9	7	7.2	0	0	51.84	0
Sum	9	35	0	0	3	102.8	10
Mean	1.8	7					

Then,

$$r = \frac{\sum_i^5 (X_i - \overline{X})(Y_i - \overline{Y})}{\sqrt{\sum_{i=1}^5 (X_i - \overline{X})^2 \sum_{i=1}^5 (Y_i - \overline{Y})^2}}$$

$$= \frac{3}{\sqrt{(102.8)(10)}} = 0.093567$$

(b)

Serial No.	X_i	Y_i	$X_i Y_i$	X_i^2	Y_i^2
1	−2	5	−10	4	25
2	−3	9	−27	9	81
3	0	6	0	0	36
4	5	8	40	25	64
5	9	7	63	81	49
Sum	9	35	66	119	255

Using the short formula:

$$r = \frac{\sum X_i Y_i - \frac{\sum X_i \sum Y_i}{n}}{\left[\left(\sum X_i^2 - \frac{(\sum X_i)^2}{n} \right) \left(\sum Y_i^2 - \frac{(\sum Y_i)^2}{n} \right) \right]^{\frac{1}{2}}}$$

$$= \frac{66 - \frac{(9)(35)}{5}}{\sqrt{\left(119 - \frac{(9)^2}{5} \right) \left(255 - \frac{(35)^2}{5} \right)}}$$

$$= \frac{3}{32.0624} = 0.093567$$

Interpretation: In general, if r > 0, X and Y move together; and if r < 0, X and Y move in opposite directions.

 More specifically, if

   (i) r = −1, we have perfect negative linear association between X and Y;

   (ii) r = 0, no linear association between X and Y;

   (iii) r = 1, we have perfect positive linear association between X and Y.

55. Using the data and results in Exercise 53 above,

  (a) Compute the slope value, b = $\left[\sum_{i=1}^5 (X_i - \overline{X})(Y_i - \overline{Y}) \right] / \left[\sum_{i=1}^5 (X_i - \overline{X})^2 \right]$, and the intercept value, a = $\overline{Y} - b\overline{X}$, of the linear equation Y = a + bX. Use this estimated equation to predict the value of Y when X = 11.

  (b) Also, reevaluate the slope value b using the following "short formula":

$$b = \frac{\sum X_i Y_i - \frac{\sum X_i \sum Y_i}{n}}{\left[\sum X_i^2 - \frac{(\sum X_i)^2}{n} \right]}$$

SOLUTION:

55. (a)

Serial No.	X_i	Y_i	$(X_i - \overline{X})$	$(Y_i - \overline{Y})$	$(X_i - \overline{X}) \times$ $(Y_i - \overline{Y})$	$(X_i - \overline{X})^2$	$(Y_i - \overline{Y})^2$
1	-2	5	-3.8	-2	7.6	14.44	4
2	-3	9	-4.8	2	-9.6	23.04	4
3	0	6	-1.8	-1	1.8	3.24	1
4	5	8	3.2	1	3.2	10.24	1
5	9	7	7.2	0	0	51.84	0
Sum	9	35	0	0	3	102.8	10
Mean	1.8	7					

Using the values from the table above, we find

$$b = \frac{\sum_{i=1}^{5}(X_i - \overline{X})(Y_i - \overline{Y})}{\sum_{i=1}^{5}(X_i - \overline{X})^2} = \frac{3}{102.8} = 0.029183$$

and

$$a = \overline{y} - b\overline{x}$$
$$= 7 - 0.029183 \times 1.8 = 6.947471. \text{ Therefore,}$$
$$Y = 6.947471 + 0.029183(11) = 7.268484.$$

(b)

Serial No.	X_i	Y_i	X_iY_i	X_i^2	Y_i^2
1	-2	5	-10	4	25
2	-3	9	-27	9	81
3	0	6	0	0	36
4	5	8	40	25	64
5	9	7	63	81	49
Sum	9	35	66	119	255

Using the values from the table above, we find

$$b = \frac{\sum X_iY_i - \dfrac{\sum X_i \sum Y_i}{n}}{\left[\sum X_i^2 - \dfrac{(\sum X_i)^2}{n}\right]}$$

$$= \frac{66 - \left(\dfrac{9 \times 35}{5}\right)}{\left(119 - \dfrac{9^2}{5}\right)} = 0.029183$$

57. Find the derivative of $y = f(x)g(x)$, where

    (a) $f(x) = 9x + 7$, $g(x) = 2x^3 + x$

    (b) $f(x) = 4x^2 - 3x$, $g(x) = x^7 - 7$

SOLUTION:

57. **(a)** $\dfrac{dy}{dx} = 9(2x^3 + x) + (9x + 7)(6x^2)$

        $= 18x^3 + 9x + 54x^3 + 42x^2$

        $= 72x^3 + 42x^2 + 9x$

    (b) $\dfrac{dy}{dx} = (8x - 3)(x^7 - 7) + (4x^2 - 3x)(7x^6)$

        $= 8x^8 - 56x - 3x^7 + 21 + 28x^8 - 21x^7$

        $= 36x^8 - 24x^7 - 56x + 21$

59. Find the derivative of $y = \dfrac{f(x)}{g(x)}$, where

    (a) $f(x) = x^2 - 9$, $g(x) = x + 3$

    (b) $f(x) = \dfrac{1}{x} + x$, $x \neq 0$, $g(x) = 2x^3 + 3$

SOLUTION:

59. **(a)** $\dfrac{dy}{dx} = \dfrac{(x + 3)(2x) - (x^2 - 9)(1)}{(x + 3)^2}$

        $= \dfrac{2x^2 + 6x - x^2 + 9}{x^2 + 6x + 9}$

    (b) $\dfrac{dy}{dx} = \dfrac{(2x^3 + 3)(-(1/x^2) + 1) - ((1/x) + x)(6x^2)}{(2x^3 + 3)^2}$

        $= \dfrac{-2x + 2x^3 - (3/x^2) + 3 - 6x - 6x^3}{4x^6 + 12x^3 + 9}$

        $= \dfrac{-4x^3 - 3x^{-2} - 8x + 3}{4x^6 + 12x^3 + 9}$

EXCEL APPLICATIONS

61. Answer Exercise 53, both parts (a) and (b), by formulating appropriate Excel spread sheets.

SOLUTION:

(a)

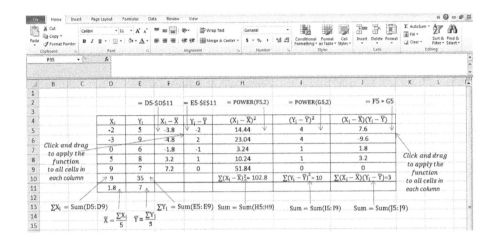

$$r = \sum_i^5 \frac{(X_i - \overline{X})(Y_i - \overline{Y})}{\sqrt{\sum_{i=1}^5 (X_i - \overline{X})^2 \sum_{i=1}^5 (Y_i - \overline{Y})^2}}$$

$$= \frac{3}{\sqrt{(102.8)(10)}} = 0.093567$$

(b)

[Excel screenshot showing the following table and formulas:]

	= D5 ∗ E5	=POWER(D5,2)	=POWER(E5,2)	
X_i	Y_i	$X_i Y_i$	X_i^2	Y_i^2
-2	5	-10	4	25
-3	9	-27	9	81
0	6	0	0	36
5	8	40	25	64
9	7	63	81	49
9	35	66	119	255

$\sum X_i = \text{Sum}(D5:D9)$

$\sum Y_i = \text{Sum}(E5:E9)$ $\sum X_i^2 = \text{Sum}(G5:G9)$

$\sum X_i Y_i = \text{Sum}(F5:F9)$ $\sum Y_i^2 = \text{Sum}(H5:H9)$

Click and drag to apply the function to all cells in each column

$$r = \cfrac{\sum X_i Y_i - \cfrac{\sum X_i \sum Y_i}{n}}{\left[\left(\sum X_i^2 - \cfrac{\left(\sum X_i\right)^2}{n}\right)\left(\sum Y_i^2 - \cfrac{\left(\sum Y_i\right)^2}{n}\right)\right]^{\frac{1}{2}}}$$

$$= \cfrac{66 - \cfrac{(9)(35)}{5}}{\sqrt{\left(119 - \cfrac{(9)^2}{5}\right)\left(255 - \cfrac{(35)^2}{5}\right)}}$$

$$= \frac{3}{32.0624} = 0.093567$$

63. Solve the simultaneous equations given in Exercises 4, 6, and 8 (a) using the Solver Add-in in Excel.

   Note: Exercises 4, 6, and 8 (a) are typed here as they are even numbers in the textbook.

   Solve the following sets of simultaneous equations:

4. $7X + 3Y = 14$
   $X - Y = 12$

6. $5X - 7Y = 28$
   $2X + 5Y = 19$

8. (a) $3P + Q = 26$
   $P - 5Q = 2$

   (b) $-6X + 3Y = 0$
   $3X + Y = 0$

   (c) $P + 2Q = 10$
   $4P + 8Q = 20$

SOLUTION:

63. Solver solution for Exercise 4.

   Let $7X + 3Y = 14$ be equation (1) and $X - Y = 12$ be equation (2).

INSTRUCTIONS:

A. Excel spread sheet set up of the problem:

   Enter data in Excel spread sheet.

   a. Type "Variables" in A1, X in B1, and Y in C1. Type "Trial values" in A2,1 in B2, and 1 in C2.

   b. Type "equation (1)" in A3 and enter the coefficient, 7, of the variable X in B3 and 3 of the variable Y in C3. In the cell D3 enter the Array formula "=Sumproduct (highlight the trial values 1 and 1, highlight 7 and 3)" and hit Enter. You will see 10 in D3, which is the left-hand side value of the equation (1) with the given trial values. In the cell E3 type "equals" (not the symbol = since Excel expects a formula to follow = symbol) and enter 14 in cell F3, which is the right-hand side value of equation (1).

c. Repeat the above step after typing equation (2) in cell A4. This will result in 0 in D4, the word "equals" in E4 and 12 in F4.

d. Click any empty cell away from the setup table, which is as follows:

	A	B	C	D	E	F
1	Variables	X	Y			
2	Trial values	1	1			
3	Equation 1	7	3	10	equals	14
4	Equation 2	1	-1	0	equals	12
5						
6						
7						

B. Using Solver Add-In*

Click on "DATA" in the menu bar and then click on "Solver" in the extreme right side of the ribbon at the top of your screen to get Solver parameters dialog box.

a. In the Solver parameters dialog box, at "set objective", select Target Cell D3, where 10 is obtained by formula.

b. At "to" select the button "value of" and enter 14 in white box.

c. Under "By Changing Variable Cells' click in the white box and highlight both the trial value cells at the same time.

d. Click inside the lower bigger rectangular box and click on "Add" on the right-hand side to get "Add Constraint" dialog box.

e. Under Cell Reference: Highlight Cell, D4 where "0" is displayed by the formula.

f. Change the middle symbol to "=".

g. Constraint: Highlight Cell F4 where "12" is typed.

h. Click "OK" since you don't have a third constraint equation.

i. Uncheck "Make Unconstrained Variables Non-Negative" box and select solving method as "Simplex LP."

j. Click on "Solve" to see a circle marked "Keep Solver Solution" in the Solver Results dialog box.

k. Click OK to see the solution as in the table below:

	A	B	C	D	E	F
1	Variables	X	Y			
2	Trial values	5	-7			
3	Equation 1	7	3	14	equals	14
4	Equation 2	1	-1	12	equals	12
5						
6						
7						

Note: The trial values of variables X and Y are changed by solver program to $X = 5$ in cell B2 and $Y = -7$ in cell C2, which satisfy both the equations. Therefore, the final solution of the two equations is $X = 5$ and $Y = -7$. It may be observed that you obtained the same solution by the elimination method in Exercise 4 above.

* If you don't find "solver" when you click on "Data" in the menu bar in Microsoft Excel (Office 2007 and onward), then

1. Click **the MICROSOFT ICON or FILE** (at the top left of your screen).
2. Click **Excel Options** (at the bottom of the menu).
3. Click **ADD-INS.**
4. Click **GO** (at the bottom of the window).
5. Check the box with "Solver Add-in."
6. Click on **OK** and see **"Solver"** at the top right of your screen.

63. Solver Solution for Exercise 6

Let $5X - 7Y = 28$ be equation (1) and $2X + 5Y = 19$ be equation (2).

INSTRUCTIONS:

A. Excel spreadsheet setup of the problem:
Enter data in Excel spreadsheet.

   a. Type "Variables" in A1, X in B1, and Y in C1. Type "Trial values" in A2, 1 in B2, and 1 in C2.

   b. Type "equation (1) " in A3 and enter the coefficient, 5, of the variable X in B3 and −7 of the variable Y in C3. In the cell D3, enter the Array formula "=Sumproduct(highlight the trial values 1 and 1, highlight 5 and −7)" and hit Enter. You will see −2 in D3, which is the left-hand side value of the equation (1) with the given trial values. In the cell E3 type "equals" (not the symbol = since Excel expects a formula to follow = symbol) and enter 28 in cell F3, which is the right-hand side value of equation (1).

   c. Repeat the above step after typing equation (2) in cell A4. This will result in 7 in D4, the word "equals" in E4 and 19 in F4.

   d. Click any empty cell away from the setup table, which is as follows:

	A	B	C	D	E	F	G
1	Variables	X	Y				
2	Trial values	1	1				
3	Equation 1	5	-7	-2	equals	28	
4	Equation 2	2	5	7	equals	19	
5							
6							
7							

B. Using Solver Add-In*
Click on "DATA" in the menu bar and then click on "Solver" on the extreme right side of your screen to get Solver parameters dialog box.

   a. In the Solver parameters dialog box, at "set objective," select Target Cell D3, where −2 is obtained by formula.

   b. At "to" select the button "value of " and enter 28 in white box.

   c. Under "By Changing Variable Cells" click in the white box and highlight both the trial value cells at the same time.

   d. Click inside the lower bigger rectangular box and click on "Add" on the right-hand side to get "Add Constraint" dialog box.

e. Under Cell Reference: Highlight Cell, D4 where "7" is displayed by the formula.
f. Change the middle symbol to "=".
g. Constraint: Highlight Cell F4 where "19" is typed.
h. Click "OK" since you don't have a third constraint equation.
i. Uncheck "Make Unconstrained Variables Non-Negative" box and select solving method as "Simplex LP."
j. Click on "Solve" to see a circle marked "Keep Solver Solution" in the Solver Results dialog box.
k. Click OK to see the solution as in the table below:

	A	B	C	D	E	F
1	Variables	X	Y			
2	Trial values	7	1			
3	Equation 1	5	-7	28	equals	28
4	Equation 2	2	5	19	equals	19
5						
6						
7						
8						

Note: The trial values of variables X and Y are changed by solver program to $X = 7$ in cell B2 and $Y = 1$ in cell C2, which satisfy both the equations. Therefore, the final solution of the two equations is $X = 7$ and $Y = 1$. It may be observed that you obtained the same solution by the elimination method in Exercise 6 above.

* If you don't find "solver" when you click on "Data" in the menu bar in Microsoft Excel (Office 2007 and onward), then

1. Click *the MICROSOFT ICON* or *FILE* (at the top left of your screen).
2. Click *Excel Options* (at the bottom of the menu).
3. Click *ADD-INS*.
4. Click *GO* (at the bottom of the window).
5. Check the box with "Solver Add-in."
6. Click on *OK* and see *"Solver"* at the top right of your screen.

63. Solver Solution for Exercise 8 (a)
    Let $3P + Q = 26$ be equation (1) and $P - 5Q = 2$ be equation (2).

INSTRUCTIONS:

A. Excel spreadsheet setup of the problem:
   Enter data in Excel spreadsheet.
   a. Type "Variables" in A1, P in B1, and Q in C1. Type "Trial values" in A2, 1 in B2, and 1 in C2.
   b. Type "equation (1)" in A3 and enter the coefficient, 3 of the variable P in B3, and 1 of the variable Q in C3. In the cell D3, enter the Array formula "=Sumproduct (highlight the trial values 1 and 1, highlight 3 and 1)" and hit Enter. You will see 4 in D3, which is the left-hand side value of the equation (1) with the given trial values. In the cell E3 type "equals" (not the symbol $=$ since Excel expects a formula to follow $=$ symbol) and enter 26 in cell F3, which is the right-hand side value of equation (1).

c. Repeat the above step after typing equation (2) in cell A4. This will result in −4 in D4, the word "equals" in E4 and 2 in F4.

d. Click any empty cell away from the setup table, which is as follows:

	A	B	C	D	E	F
1	Variables	P	Q			
2	Trial values	1	1			
3	Equation 1	3	1	4	equals	26
4	Equation 2	1	-5	-4	equals	2
5						
6						

B. Using Solver Add-In*

Click on "DATA" in the menu bar and then click on "Solver" in the extreme right side of the ribbon at the top of your screen to get Solver parameters dialog box.

a. In the Solver parameters dialog box, at "set objective," select Target Cell D3, where 4 is obtained by formula.

b. At "to" select the button "value of" and enter 26 in white box.

c. Under "By Changing Variable Cells" click in the white box and highlight both the trial value cells at the same time.

d. Click inside the lower bigger rectangular box and click on "Add" on the right-hand side to get "Add Constraint" dialog box.

e. Under Cell Reference: Highlight Cell, D4 where "−4" is displayed by the formula.

f. Change the middle symbol to "=".

g. Constraint: Highlight Cell F4 where "2" is typed.

h. Click "OK" since you don't have a third constraint equation.

i. Uncheck "Make Unconstrained Variables Non-Negative" box and select solving method as "Simplex LP."

j. Click on "Solve" to see a circle marked "Keep Solver Solution" in the Solver Results dialog box.

k. Click OK to see the solution as in the table below:

	A	B	C	D	E	F
1	Variables	P	Q			
2	Trial values	8.25	1.25			
3	Equation 1	3	1	26	equals	26
4	Equation 2	1	-5	2	equals	2
5						

Note: The trial values of variables P and Q are changed by solver program to $P = 8.25$ in cell B2 and $Q = 1.25$ in cell C2, which satisfy both the equations. Therefore, the final solution of the two equations is $P = 8.25$ and $Q = 1.25$. It may be observed that you obtained the same solution by the elimination method in Exercise 8 (a) above.

* If you don't find "solver" when you click on "Data" in the menu bar in Microsoft Excel (Office 2007 and onward), then

1. Click *the MICROSOFT ICON or FILE* (at the top left of your screen).

2. Click *Excel Options* (at the bottom of the menu).

3. Click *ADD-INS*.

4. Click *GO* (at the bottom of the window).

5. Check the box with "Solver Add-in."

6. Click on *OK* and see "Solver" at the top right of your screen.

APPENDIX A EXERCISES

Simplify the following:

A.1 $(18 \div 12) \times 8^2 - 88$

A.3 $[(4)^2 \times 4] \times 16 + 10$

A.5 $X^2 Y X Y^3 X^2 Y^2$

A.7 $\left(\dfrac{X^3}{-64Y^6} \right)^{1/3}$

A.9 $\left[3\sqrt{\dfrac{1}{27}} \right]^2$

A.11 $0.30 + 1.645 \sqrt{\dfrac{0.3(1 - 0.3)}{100}}$

A.13 $\{(0.3)xyz\}^0$

A.15 $\dfrac{(0.3)^4}{(0.2)^3} \times 10^3$

A.17 $\left(\dfrac{1.24(0.25)}{0.62} \right)^{1/2}$

A.19 $x \cdot x \cdot y^2$

A.21 $\left(x^4 \right)^4$

A.23 $(4 - 2^2)^0$

A.25 $\left(\dfrac{x^3}{y^4} \right)^5$

A.27 $\left(\dfrac{x^4}{x^3} \right)^5$

A.29 $\dfrac{12^3}{12^0}$

A.31 $\left(\dfrac{9}{16} \right)^{1/2}$

A.33 $27^{4/3}$

A.35 $\dfrac{x^8 / y^4}{8 / y^2}$

A.37 $\left(\dfrac{1}{3} \right)^3$

SOLUTION:

A.1. $\dfrac{18}{12} \times 64 - 88 = \dfrac{3}{2} \times 64 - 88 = 96 - 88 = 8$

SOLUTION:

A.3 $[16 \times 4] \times 16 + 10 = 64 \times 16 + 10 = 1024 + 10 = 1034$

SOLUTION:

A.5 $x^{2+1+2}y^{2+3+1} = x^5 y^6$

SOLUTION:

A.7 $\left(\dfrac{x^{3/3}}{-4^{3/3}y^{6/3}} \right) = \dfrac{x}{-4y^2}$

SOLUTION:

A.9 $\left[\dfrac{1}{3^{3/3}} \right]^2 = \dfrac{1}{9}$

SOLUTION:

A.11 $0.30 + 1.645\sqrt{\dfrac{0.3 \times 0.7}{100}}$

$0.30 + 1.645\sqrt{\dfrac{0.21}{10^2}} = 0.30 + 1.645\sqrt{0.0021} = 0.30 + 0.075 = 0.375$

SOLUTION:

A.13 $(0.3)^0 \times (xyz)^0 = 1 \times 1 = 1$

SOLUTION:

A.15 $\dfrac{0.3 \times 0.3 \times 0.3 \times 0.3}{0.2 \times 0.2 \times 0.2} \times 10^3 = \dfrac{0.0081}{0.008} \times 1000 = 1012.5$

SOLUTION:

A.17 $\left(\dfrac{0.31}{0.62} \right)^{1/2} = \sqrt[2]{0.5} = 0.7071$

SOLUTION:

A.19 $x^{1+1}y^2 = x^2 y^2$

SOLUTION:

A.21 $x^4 \times x^4 \times x^4 \times x^4 = x^{4+4+4+4} = x^{16}$

SOLUTION:

A.23 $(4 - 4)^0 = 0^0 =$ undefined

SOLUTION:

A.25 $\dfrac{x^{15}}{y^{20}} = x^{15} \times y^{-20}$

SOLUTION:

A.27 $\left(\dfrac{x^{4\times5}}{x^{3\times5}}\right) = \dfrac{x^{20}}{x^{15}} = x^{20-15} = x^5$

SOLUTION:

A.29 $12^{3-0} = 12^3 = 1728$

SOLUTION:

A.31 $\dfrac{\sqrt[2]{9}}{\sqrt[2]{16}} = \dfrac{3}{4}$

SOLUTION:

A.33 $3^{3\times(4/3)} = 3^4 = 81$

SOLUTION:

A.35 $\dfrac{x^8}{y^4} \times \dfrac{y^2}{8} = \dfrac{x^8}{y^{4-2} \times 8} = \dfrac{x^8}{8y^2}$

SOLUTION:

A.37 $\dfrac{1}{3} \times \dfrac{1}{3} \times \dfrac{1}{3} = \dfrac{1}{27}$

SOLUTION:

A.39 Evaluate $y = \left(\dfrac{1}{x}\right)^{-(4/6)}$ at $x = 27$

$$Y = \left(\dfrac{1}{27}\right)^{-(4/6)} = \left(\dfrac{1}{(3)^{3(-4/6)}}\right) = \dfrac{1}{3^{-2}} = 3^2 = 9$$

Factorize the following:

A.41 $x^2 - 7x + 12$

A.43 $x^2 + x - 12$

A.45 $5x^2 + 7x + 2$

A.47 $x^2 + 5x + 6$

SOLUTION:

A.41 $x^2 - 3x - 4x + 12 = x(x - 3) - 4(x - 3) = (x - 3)(x - 4)$

SOLUTION:

A.43 $x^2 + 4x - 3x - 12 = x(x + 4) - 3(x + 4) = (x + 4)(x - 3)$

SOLUTION:

A.45 $5x^2 + 5x + 2x + 2 = 5x(x + 1) + 2(x + 1) = (5x + 2)(x + 1)$

SOLUTION:

A.47 $x^2 + 2x + 3x + 6 = x(x + 2) + 3(x + 2) = (x + 3)(x + 2)$

Evaluate the following expressions:

A.49 $\dfrac{7}{8} - \dfrac{3}{24}$

A.51 $\dfrac{1}{2} - \dfrac{1}{4}$

SOLUTION:

A.49 $LCM = 24 = 2 \times 2 \times 2 \times 3$

$$\frac{21}{24} - \frac{3}{24} = \frac{21 - 3}{24} = \frac{18}{24} = \frac{3}{4}$$

SOLUTION:

A.51 $LCM = 4 = 2 \times 2$

$$\frac{2}{4} - \frac{1}{4} = \frac{2 - 1}{4} = \frac{1}{4}$$

Solve the following:

A.53 $\dfrac{(1/3) - (1/5)}{(2/3) - (1/6)}$

A.55 $\dfrac{3}{5} \times \dfrac{-2}{7}$

A.57 $\dfrac{-2}{3} + \dfrac{3}{5}$

A.59 $\dfrac{0}{5} + \dfrac{5}{6}$

SOLUTION:

A.53 $LCM\,1 = 5 \times 3 = 15$

$\quad\quad\quad LCM\,2 = 2 \times 3 = 6$

$$\frac{(5/15) - (3/15)}{(4/6) - (1/6)} = \frac{(5 - 3)/15}{(4 - 1)/6} = \frac{2/15}{3/6} = \frac{2}{15} \times \frac{6}{3} = \frac{4}{15}$$

SOLUTION:

A.55 $\dfrac{3 \times -2}{5 \times 7} = \dfrac{-6}{35}$

SOLUTION:

A.57 $\dfrac{-2}{3} \times \dfrac{5}{3} = \dfrac{-2 \times 5}{3 \times 3} = \dfrac{-10}{9}$

SOLUTION:

A.59 $\left(0 \times \dfrac{6}{5}\right) = 0$

Simplify the following:

A.61 $2\dfrac{3}{5} + \dfrac{1}{3} - \dfrac{2}{7}$

A.63 $\left(2\dfrac{3}{4}\right)\left(4\dfrac{1}{2}\right)$

A.65 $\left(2\dfrac{3}{4}\right) / \left(4\dfrac{1}{3}\right)$

SOLUTION:

A.61 $\dfrac{13}{5} + \dfrac{1}{3} - \dfrac{2}{7}$

LCM $= 3 \times 5 \times 7$

$\dfrac{273}{105} + \dfrac{35}{105} - \dfrac{30}{105} = \dfrac{273 + 35 - 30}{105} = \dfrac{278}{105}$

SOLUTION:

A.63 $\left(\dfrac{11}{4}\right)\left(\dfrac{9}{2}\right) = \dfrac{99}{8}$

SOLUTION:

A.65 $\dfrac{11}{4} \times \dfrac{3}{13} = \dfrac{33}{52}$

Show the position of each digit in the following:

A.67 0.873

SOLUTION:

A.67 $\dfrac{8}{10} + \dfrac{7}{100} + \dfrac{3}{1000} = 0.8 + 0.07 + 0.003$

Find:

A.69 $1.273 + 3.2$

A.71 $0.34 - 0.07 + 1.7$

SOLUTION:

A.69 4.473

SOLUTION:

A.71 $0.34 + 1.7 - 0.07 = 1.97$

Evaluate:

A.73 $0.1 \times 0.01 \times 0.005$

A.75 $\dfrac{(1.5)(-0.6) - (2.5)(1.7)}{20}$

A.77 $\dfrac{0.4}{0.04} - \dfrac{6(3.6)}{0.072}$

SOLUTION:

A.73 0.000005

SOLUTION:

A.75 $\dfrac{-0.9 - 4.25}{20} = -0.2575$

SOLUTION:

A.77 $10 - \dfrac{21.6}{0.072} = 10 - 300 = -290$

Round-off the resulting decimal to the given place:

A.79 $\dfrac{5}{7}$, to the nearest hundredths

A.81 $\dfrac{22}{7}$, to the nearest ten thousandth

SOLUTION:

A.79 0.71

SOLUTION:

A.81 3.1429

A.83 In a business school, the ratio of female to male undergraduate students is 1 to 4. After 200 female students are admitted, the ratio of female to male students became 2/3. Find the total number of students after admitting 200 female students in the school.

SOLUTION:

A.83 Let x denote number of female students to start with.
Then,

$$\frac{x + 200}{4x} = \frac{2}{3}$$

$$3x + 600 = 8x$$

$$5x = 600$$

$$x = \frac{600}{5}$$

$$x = 120$$

So, male students to start with are 480. After 200 female students joined, the total number of female students is $120 + 200 = 320$.

Therefore, the total number of students after the 200 female students are admitted is $320 + 480 = 800$.

A.85   Mr. John Smith receives $1400.00 as salary every week after taxes. If he is in the 30% tax bracket, what is his salary per week before taxes?

SOLUTION:

A.85   Let's call his salary before Taxes "S."

$$100\% - 30\% = 70\%$$

70% of Mr. John Smith's salary before taxes (S) is equal to 1400.00.

$$\frac{70}{100} \times S = 1400, \qquad S = \frac{1400 \times 100}{70} = 2000 \text{ (which is in dollars)}$$

$$S = \$2000.00$$

A.87   Fill in the blanks:

Row	Fraction	Decimal	%, Percentage
1	---------	0.02	--------
2	---------	0.125	---------
3	----------	3.750	---------
4	$\dfrac{1}{16}$	----------	----------
5	$\dfrac{1}{8}$	----------	-----------
6	$\dfrac{3}{2}$	-----------	-----------
7	-----------	-----------	30
8	-----------	-----------	50
9	-----------	-----------	10
10	-----------	-----------	200
11	$\left(\dfrac{1}{4}\right)^2$	-----------	-----------
12	$\left(\dfrac{1}{10}\right)^5$	------------	-----------

SOLUTION:

A.87

Row	Fraction	Decimal	%, Percentage
1	$\dfrac{1}{50}$	0.02	2
2	$\dfrac{1}{8}$	0.125	12.5
3	$\dfrac{15}{4}$	3.750	375
4	$\dfrac{1}{16}$	0.0625	6.25
5	$\dfrac{1}{8}$	0.125	12.5
6	$\dfrac{3}{2}$	1.5	150
7	$\dfrac{3}{10}$	0.3	30
8	$\dfrac{1}{2}$	0.5	50
9	$\dfrac{1}{10}$	0.1	10
10	2	2.0	200
11	$\left(\dfrac{1}{4}\right)^2$	0.0625	6.25
12	$\left(\dfrac{1}{10}\right)^5$	0.00001	0.001

A.89 *The Wall Street Journal*, Friday, May 25, 2012, reported that the 52-week Crude Oil high and low prices per barrel were, respectively, \$109.77 and \$75.67. Find the percentage decrease from high price to low price.

SOLUTION:

A.89 Percentage change is

$$\frac{109.77 - 75.67}{109.77} \times 100 = \frac{34.1}{109.77} \times 100\% = 0.3106 \times 100\% = 31.06\%$$

Chapter 2

Applications of Linear and Nonlinear Functions: A Summary

2.2 LINEAR DEMAND AND SUPPLY FUNCTIONS

Let us specify the *linear demand function* for a product (or service) as $Q = a + bP + cY$, where Q is the quantity demanded, P is the product price, and Y is the income. Here, $b < 0$—since a demand function is downward sloping with respect to price—(Figure 2.1a) and $c > 0$—demand increases with income (Figure 2.1b).

A particular type of demand function is an aggregate demand function that is generally called a *consumption function* and expressed as $C = a + bY$, where C is the consumption expenditure, Y is the disposable income, a is the subsistence level of income when $Y = 0$, and $b = \Delta C / \Delta Y$ (the slope) is the *marginal propensity to consume*. The break-even level of income occurs where $C = Y$ (point A in Figure 2.2). Note that positive saving occurs when $Y > Y_A$; dissaving occurs when $Y < Y_A$.

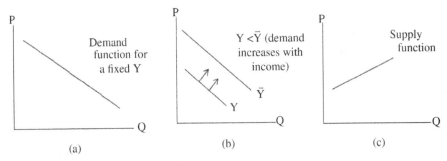

Figure 2.1 Demand and supply functions.

Solutions Manual to Accompany Introduction to Quantitative Methods in Business: With Applications Using Microsoft® Office Excel®, First Edition. Bharat Kolluri, Michael J. Panik, and Rao N. Singamsetti.
© 2017 John Wiley & Sons, Inc. Published 2017 by John Wiley & Sons, Inc.

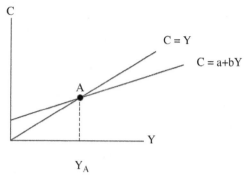

Figure 2.2   Consumption function.

A linear supply function for a product can be written as $Q = c + dP$, where Q is the quantity supplied and P is the product price. Here, $d > 0$ since supply functions are upward sloping (Figure 2.1c).

2.3   LINEAR TOTAL COST AND TOTAL REVENUE FUNCTIONS

The *total cost function* is defined as $TC = VC \times Q + FC$, where Q is the quantity, VC is the unit variable cost (thus, $VC \times Q$ is the *total variable cost*), and FC is the *fixed cost*. The *total revenue function* is written as $TR = P \times Q$ (Figure 2.3). The break-even point A occurs where $TR = TC$. *Total profit* is defined as $P = TR - TC$. Clearly, $P = 0$ at the break-even quantity Q_E. For $Q > Q_E$, profit is positive; for $Q < Q_E$, the firm incurs a loss or profit is negative.

2.4   MARKET EQUILIBRIUM

Given the demand and supply functions $Q_D = f(P_D)$ and $Q_S = g(P_S)$, respectively, a *market equilibrium* is a point where the quantity demanded and the quantity supplied are equal. We shall write the *equilibrium condition*—a condition that aligns

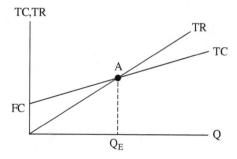

Figure 2.3   Break-even output.

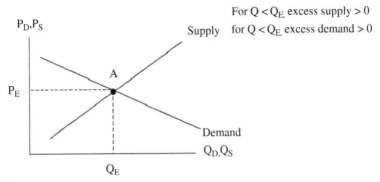

Figure 2.4   Market equilibrium.

quantity demanded with quantity supplied—as simply $Q_D = Q_S$. Hence, the complete market model appears as

$$Q_D = f(P_D)$$
$$Q_S = g(P_S)$$
$$Q_D = Q_S$$

and is illustrated in Figure 2.4. (Note that when $Q_D = Q_S$, it follows that $P_D = P_S$.)

2.6   APPLICATIONS OF NONLINEAR FUNCTIONS

If an initial amount P_0 grows at a rate of R% per annum for t periods, the accumulated value is

$$P_t = P_0\left(1 + \frac{R}{n}\right)^{nt},$$

where n is the number of times the increment is added per period, that is, it is the frequency of compounding within each period. For example, if

a. $n = 1$ (yearly compounding),

$$P_t = P_0(1 + R)^t;$$

b. $n = 2$ (semiannual compounding),

$$P_t = P_0\left(1 + \frac{R}{2}\right)^{2t};$$

c. $n = 4$ (quarterly compounding),

$$P_t = P_0\left(1 + \frac{R}{4}\right)^{4t};$$

d. $n = 12$ (monthly compounding),

$$P_t = P_0 \left(1 + \frac{R}{12}\right)^{12t}.$$

As n approaches infinity, compounding is continuous so that P accumulates over time according to the expression

$$P_t = P_0 e^{rt}, \quad e \cong 2.71828.$$

This continuous compounding equation is often called an *exponential growth function*.

2.7   PRESENT VALUE OF AN INCOME STREAM

Suppose f (t) is an income stream obtained continuously over the period from $t = 0$ to $t = X$. The *present value of the income stream*, f (t), compounded continuously at an interest rate of r% per year over the time interval [0, X], where X denotes the number of years the income stream lasts, is

$$P = \int_0^x f(t) e^{-rt}.$$

For instance, if $r = 0.02$ (2% per annum), $X = 3$, and f (t) = $10,000 per year, then the present value of this income stream is

$$\begin{aligned}
P &= \int_0^3 10,000 \, e^{-0.02t} dt \\
&= 10,000 \left(\frac{1}{-0.02}\right) e^{-0.02t} \Big]_{t=0}^{t=3} \\
&= -500,000 \left(e^{-0.02(3)} - e^{-0.02(0)}\right) \\
&= -500,000(0.94176 - 1) \\
&= \$29,117.73.
\end{aligned}$$

Thus, the present value of $10,000 per year for 3 years at a 2% per annum interest rate is about $29,118.00.

2.8   AVERAGE VALUES

In general, if $y = f(x)$, then the *average value of* y is calculated as

$$\frac{y}{x} = \frac{f(x)}{x}.$$

More specifically:

$$\text{Average Cost (AC)} = \frac{TC}{Q} = \frac{VC \times Q + FC}{Q}$$

$$= VC + \frac{FC}{Q}$$

$$= \text{average variable cost} + \text{average fixed cost}.$$

$$\text{Average Revenue (AR)} = \frac{TR}{Q} = \frac{P \times Q}{Q} = P = \text{price}.$$

Note: If P is not constant but determined from the (inverse) demand function as $P(Q) = \alpha + \beta Q$, then

$$AR = \frac{(\alpha + \beta Q)Q}{Q} = \alpha + \beta Q = P(Q) = \text{price}.$$

2.9 MARGINAL VALUES

In general, if $y = f(x)$, then the *marginal value of* y with respect to x is defined as

$$\frac{dy}{dx} = f'(x) = \text{slope of f at a given x}.$$

More specifically:

Marginal cost (MC) $= \dfrac{dC(Q)}{dQ}$, where C(Q) is the total cost function;

Marginal revenue (MR) $= \dfrac{dR(Q)}{dQ}$, where R(Q) is the total revenue function.

2.10 ELASTICITY

The *point elasticity* E of $y = f(x)$ at a given x is defined as the ratio of a percentage change in y to a percentage change in x or

$$E = \frac{dy/y}{dx/x} = \frac{x}{y}\frac{dy}{dx} = \frac{d \log y}{d \log x}.$$

For instance, the *price elasticity of demand* at a given price—quantity combination is calculated as

$$E = -\frac{dQ}{dP}\frac{P}{Q}.$$

(The minus sign appears because $(dQ/dP) < 0$, that is, the demand curve is downward sloping.) If

E > 1, demand is *elastic*;

E < 1, demand is *inelastic*;

E = 1, demand is of *unitary elasticity*.

SOLUTIONS TO ODD-NUMBERED EXERCISES

1. Determine the demand curve for the price–quantity data in the following table for Q = f(P). Also show it graphically.

P in $	0	10	20	30	40
Q in 1000's	8	6	4	2	0

SOLUTION:

1. From the data, the slope $= \dfrac{\Delta Q}{\Delta P} = \dfrac{6-8}{10-0} = \dfrac{-2}{10} = \dfrac{-1}{5}$.

Therefore, we can write the demand equation as $Q = -(-1/5)\,P + C$, where C is the intercept. Substituting one of the given points from the table above, say (0, 8), we obtain $8 = (-1/5)\,(0) + C$. Thus, $C = 8$ and the demand equation is given by $Q = (-1/5)P + 8$.
The graph of this equation is given below.

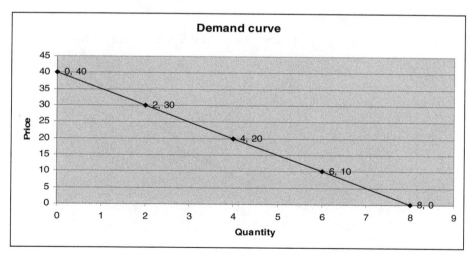

3. For the consumption function $C = 20 + (2/3)Y$:

   a. Find the break-even income level.

   b. Find the consumption expenditure at income levels of 40 and 80. (In this exercise, assume C and Y are measured in thousands of dollars.)

    c. Show the answers to (a) and (b) graphically and identify the ranges of income corresponding to dissaving and saving.

    d. Plot the *savings function* $S = Y - C$. What is the *marginal propensity to save* (MPS)? Does $MPC + MPS = 1$? At $Y = 20$, does $Y = C + S$?

SOLUTION:

3. **a.** Here, the break-even income level can be found by solving two equations, namely,

$$C = 20 + \frac{2}{3}Y, \quad \text{and} \quad C = Y.$$

Subtracting the second equation from the first, we obtain $0 = 20 + (2/3)Y - Y$ or $0 = 20 + ((2/3) - 1)Y$ or $0 = 20 + (-1/3)Y$

$$\frac{-1}{3}Y = -20.$$

Multiplying by -3 on both sides, we obtain $Y = 60$.
        Thus, the break-even level of income is 60.

    b. Consumption at $Y = 40$ is

$$C = 20 + \frac{2}{3} \times 40 = 20 + \frac{80}{3}$$

or
$$C = 20 + 26.67 = 46.67 = \$46,666.67.$$

Consumption at $Y = 80$ is
$$C = 20 + \frac{2}{3} \times 80,$$
$$C = 20 + 53.33 = 73.33 = \$73,333.33.$$

    c. Graphs for (a) and (b):

$$C = 20 + \frac{2}{3}Y \text{ and } C = Y.$$

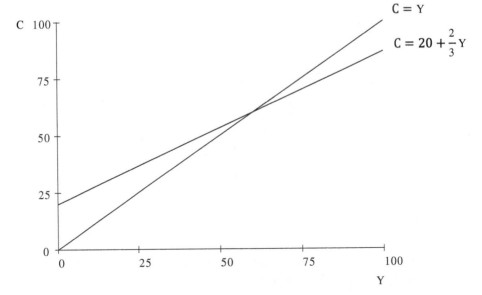

From the graph above, we can see that up to the income level of $Y = 60$, consumption exceeds income. Therefore, the income range of 0–60 corresponds to dissaving, and the income level above 60 indicates saving.

d. Savings function, $S = Y - C = Y - (20 + (2/3)Y) = -20 + (1/3)Y$.

To plot the savings function, prepare a table as follows:

Y	S
30	−10
60	0
90	10

Marginal propensity to save (MPS) is the slope of the savings function, and it is $1/3$.

Savings function

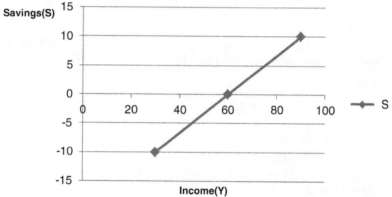

Now, MPC + MPS $= (2/3) + (1/3) = 1$. At $Y = 20$, $C = 20 + (2/3)(20) = 20 + 13.33 = 33.33$ and $S = -20 + (1/3)(20) = -20 + 6.67 = -13.33$. Hence, $C + S = 33.33 - 13.33 = 20 = Y$. Thus, at $Y = 20$, C and S add up to 20. In fact, $Y = C + S$ holds at any level of Y.

5. a. Find the break-even output level with the following information:

$$\text{Selling price} = \$5.00$$

$$\text{Unit variable cost} = \$2.00$$

$$\text{Fixed cost} = \$3000$$

b. Suppose a particular firm is producing and selling fig products in Toronto. The selling price of a 10 oz jar of fig jam is $9. Fixed cost of producing the fig jam is $540 and unit variable cost of a 10 oz jar of fig jam is $6. Find the break-even quantity.

SOLUTION:

5. a. Using the same notation as in Exercise 4, we find $TC = 3000 + 2Q$ and $TR = 5Q$.
At the break-even output level Q_E, $TR = TC$.
Thus, $5Q_E = 3000 + 2Q_E$.
$5Q_E - 2Q_E = 3000$ or $3Q_E = 3000$, and therefore $Q_E = 1000$.

b. Selling price of jam $= \$9$.
Fixed cost of producing jam $= \$540$.
To find the break-even quantity, let Q be break-even quantity.
Then, $TR = 9Q$.
$TC = 540 + 6Q$.
At the break-even level, $TR = TC$.
Therefore, $9Q = 540 + 6Q$.

$$9Q - 6Q = 540.$$

$$3Q = 540.$$

$$Q = \frac{540}{3} = 180.$$

7. In Example 2.5 in the text, find the quantity demanded and the quantity supplied if the price is artificially fixed at $8000. (Note that at this price of $8000, which is lower than the equilibrium price, the quantity demanded far exceeds the quantity supplied. Thus, there will be product shortage.)

SOLUTION:

7. At the price level of $8000, quantity demanded is given by
$80,000 = 16,000 - 20Q_D$,
$20Q_D = 16,000 - 8000$,
$20Q_D = 8000$ and $Q_D = 8000/20 = 400$.
Therefore, quantity demanded is 400 units when price is fixed at $8000.
Supply function is given as

$$P_S = 6000 + 30Q_S$$

Using the price of $8000 in the above supply function, we obtain
$8000 = 6000 + 30Q_S$.
$30Q_S = 8000 - 6000 = 2000$. Thus,
$Q_S = 2000/30 = 66.67$ or approximately 67 units.
Q_D is 400 units and Q_S is 67 units at a fixed price of $8000, resulting in an excess demand of 333 $(400 - 67)$ units. This situation implies an imbalance in supply and demand in the market and, as a result, there will be product shortage.

9. The following are the supply and demand functions for a portable music player. Draw the supply and demand graphs for the music player and find the equilibrium price and quantity.

$$Q_D = 560 - 10P_D,$$
$$Q_S = 80 + 5P_S.$$

SOLUTION:

9. Graphs of supply and demand functions:

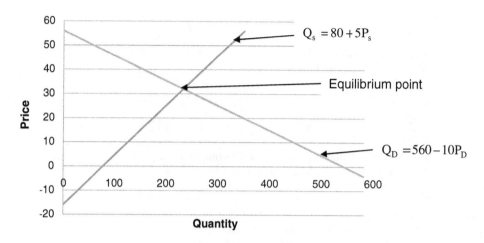

At equilibrium point, $Q_s = Q_d = Q$ and $P_s = P_d = P$.
Thus, the demand and supply equations assume the following form:

$$Q = 560 - 10P \text{ (demand)}$$
$$Q = 80 + 5P \quad \text{(supply)}$$

Since each equals Q,

$$80 + 5P = 560 - 10P$$
$$10P + 5P = 560 - 80$$
$$15P = 480$$
$$P = \$32.$$

At $P = 32$, $Q = 80 + 5(32) = 80 + 160 = 240$ units.

11. Find the future value of an investment of $1000.00 in 10 years if the interest rate of 10% per year is compounded continuously.

SOLUTION:

11. The future value of an investment of P_0 ($1000) after $t = 10$ years, if the interest rate of 10% per year is compounded continuously, is given by the equation

$$P_t = P_0 e^{0.1 \times 10} = 1000e^1 = 1000 \times 2.71828 = 2718.28.$$

13. A company offers an employee a pension plan providing payment of $60,000 per year beginning at the age of 65 along with a one-time cash payment of $100,000. Find the present value of the total amount, assuming an average interest rate of 5% per year.

SOLUTION:

13. Present value of the pension plan $= 100,000 + \int_0^t 60,000e^{-0.05x}\,dx$

$$= 100,000 + 60,000 \times \frac{1}{-0.05}\,[e^{-0.05x}]_0^t$$

This is the case of perpetuity, and therefore we assume that t tends to ∞ and, because of this, the above integral value equals

$$100,000 + \frac{60,000}{-0.05}\,[e^{-0.05t} - e^{-0.05(0)}] = 100,000 - 1,200,000[0-1]\text{ as }t \to \infty$$

$$= 100,000 + 1,200,000 = 1,300,000.$$

Here it is assumed that the interest rate is compounded continuously and the pension is paid to the retiring employee and the surviving heirs.

15. Find the savings function for an individual whose marginal savings rate is a constant at 10% of disposable income Y (income minus taxes). Assume that the amount of savings is zero when the disposable income for the person is zero but the individual needs $15,000 as minimum consumption for survival.

SOLUTION:

15. The total savings function, denoted by S(Y), is given by $\int (1/10)dY$, where Y stands for a particular value of income and dY represents change in income. The value of the above integral is $(1/10)Y + k$, where k is the constant of integration. We know that $Y = S + C$ from the result of Exercise 3(d). Substituting, $Y = (1/10)Y + k + C$. When $Y = 0$, we obtain $0 = (1/10)Y(0) + k + 15,000$, where 15,000 is assumed minimum consumption level (C). Thus, the constant of integration $k = -15,000$, which can be called dissaving. Therefore, the savings function is $S(Y) = (1/10)Y - 15,000$.

17. Suppose the marginal revenue function of an auto manufacturing company is given by MR $= 100,000 - 2Q$, where Q stands for the number of autos produced. Derive the total revenue function as well as the demand function for the autos.

SOLUTION:

17. The total revenue, denoted by TR(Q), is given by $\int (100,000 - 2Q)dQ = 100,000Q - (2Q^2/2) + c = 100,000Q - Q^2 + c$, where $c = 0$ since TR(0) $= 0$. Thus, TR(Q) $= 100,000Q - Q^2$.
The demand function $P = (TR/Q) = (100,000Q - Q^2) \div Q = 100,000 - Q$.

19. Find the average values for functions a, b, and c given below and evaluate each at Q = 100.

   a. $C(Q) = 540 + 6Q$, where C(Q) is the total cost function and Q is the level of production.

   b. $R(Q) = 9Q$, where R(Q) is the total revenue when Q units are transacted at a constant price of P = $9.

   c. $R(Q) = P(Q)Q$, where $P(Q) = 40 - 0.1Q$, is a demand (inverse) function showing the dependence of price on quantity demanded Q.

SOLUTION:

19. a. Average cost $(AC) = \dfrac{C(Q)}{Q} = \dfrac{540 + 6Q}{Q} = \dfrac{540}{Q} + 6.$

At $Q = 100$, $AC = (540/100) + 6 = 5.4 + 6 = 11.4$

b. Average revenue $(AR) = \dfrac{R(Q)}{Q} = \dfrac{9Q}{Q} = 9.$

c. Average revenue $(AR) = \dfrac{R(Q)}{Q} = \dfrac{P(Q)Q}{Q} = P(Q) = 40 - 0.1Q.$

At $Q = 100$, $AR = 40 - 0.1(100) = 40 - 10 = 30.$

21. Find the price elasticity of demand for the following demand function at price levels of \$2, \$3.33, and \$4, and interpret price elasticity at each price level. $Q = 20 - 3P$.

SOLUTION:

21. Given the demand function $Q = 20 - 3P$, $(dQ/dP) = -3$, and thus

$$E_p = = = \frac{P}{Q} \times \frac{dQ}{dP}, \quad \text{where } E_p \text{ is the price elasticity,}$$

$$E_p = \frac{2}{20 - 3(2)} \times (-3), \quad \text{when } P = 2,$$

$$= \frac{-6}{14} = -0.429$$

We can conclude that, as $|E_p| < 1$, the demand is inelastic at $P = 2$, that is, with a 1% change in price, there will be a less than 1% change in the quantity demanded. At $P = \$3.33$,

$$E_p = \frac{P}{Q} \times \frac{dQ}{dP}$$

$$= \frac{3.33}{20 - 3(3.33)} \times (-3)$$

$$= \frac{-9.99}{10.01} = -0.998 \approx -1$$

We can conclude that as $|E_p| = 1$, the demand is unit elastic at $P = 3.33$, that is, with a 1% change in price, there will be a 1% change in the quantity demanded. At $P = \$4$,

$$E_p = \frac{P}{Q} \times \frac{dQ}{dP}$$

$$= \frac{4}{20 - 3(4)} \times (-3) = \frac{-12}{8} = -1.5.$$

We can conclude that as $|E_p| > 1$, the demand is elastic at $P = 4$, that is, with a 1% change in price, there will be a more than 1% change in the quantity demanded.

23. Find the price elasticity of the supply function $Q = 11 + 7P$ at $P = 10$.

SOLUTION:

23. The supply function is $Q = 11 + 7P$.

$$\frac{dQ}{dP} = 7$$

$$E_p = \frac{P}{Q} \times \frac{dQ}{dP}$$

$$= \frac{10}{11 + 7(10)} \times 7 = \frac{70}{11 + 70}$$

$$= \frac{70}{81} = 0.86.$$

As $|E_p| < 1$, we can conclude that our supply is inelastic, that is, a 1% change in price will result in less than a 1% change in the quantity supplied.

EXCEL APPLICATIONS

25. This exercise deals with the case where price is not a constant but related to quantity demanded. Demand (price) function is $P = 50 - 2.5Q$, where Q stands for quantity demanded. The revenue function is $TR = Q \times P = 50Q - 2.5Q^2$ and the total cost function is $TC = 25 + 25Q$. Using Excel, develop a table showing columns for Q, P, TR, TC, and Profit (TR-TC). Graph the total cost and total revenue functions at the values of Q: 0, 1, 2, 3, 4, 5, 6, 7, 8, and 9. Indicate approximately on the graph the break-even output levels, where total cost equals total revenue. Distinguish between these two output levels.

SOLUTION:

25.

Q	P	TR	TC	Profit (TR-TC)
0	50	0	25	−25
1	47.5	47.5	50	−2.5
2	45.0	90	75	15
3	42.5	127.5	100	27.5
4	40	160	125	35
5	37.5	187.5	150	37.5
6	35	210	175	35
7	32.5	227.5	200	27.5
8	30	240	225	15
9	27.5	247.5	250	−2.5

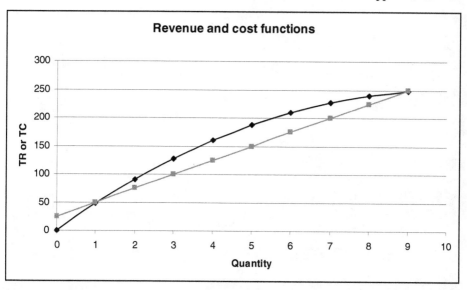

From the graph, we can see that there are two break-even points. One can show that one point is where $Q = 1.13$ and the second point is where $Q = 8.88$.

As you increase Q, at the first point, the loss will be zero and, at the second point, the profit will be zero. This is the difference between these two break-even levels of output.

27. Using Excel, draw the curve of the following function:

$$Y = (1/\sqrt{2\pi})\ \exp\left(-(1/2)\ x^2\right),$$ where $\pi = 22/7$. Consider the following values of $x = -3, -2, -1, 0, 1, 2, 3$ for preparing a table and drawing the graph.

SOLUTION:

27.

Instructions:

Step 1: Open Excel spreadsheet and enter x values in column A as shown above.

Step 2: In cell B2, enter the formula as shown in the f_x, formula bar, and hit "Enter" button.

Step 3: After preparing the table, highlight the table and select 'Insert' on the toolbar and choose the 'scatter' icon and select the curve option to get the curve as shown above.

| | B2 | ▼ | f_x | =(1/SQRT(2*(22/7))*EXP(-(A2^2)/2)) |

	A	B	C	D	E	F	G	H	I
1	x	y							
2	-4	0.000134							
3	-3	0.004431							
4	-2	0.05398							
5	-1	0.241922							
6	0	0.398862							
7	1	0.241922							
8	2	0.05398							
9	3	0.004431							
10	4	0.000134							
11									
12									

Chapter 3

Optimization: A Summary

3.2 UNCONSTRAINED OPTIMIZATION

3.2.1 Models of Profit and Revenue Maximization

For profit maximization, the optimal decision rule is that the firm should produce the level of output where $MR = MC$. Note that

a. if $P = $ constant, the optimal output level occurs where $MR = MC = P$;

b. if $P \neq $ constant, product price is determined from the (inverse) demand function $P(Q) = \alpha + \beta Q$.

For revenue maximization, the optimal decision rule is that the firm should produce the level of output where $MR = 0$.

3.2.3 Solution Using the Calculus Approach

Suppose we have a continuously differentiable function $y = f(x)$ over an open domain (a,b). Let $x_0 \in (a, b)$. Then,

1. if $f'(x_0) = 0$ and $f''(x_0) < 0$, f has a maximum at x_0;

2. if $f'(x_0) = 0$ and $f''(x_0) > 0$, f has a minimum at x_0.

3.2.5 Solution Using the Calculus Approach

Denote the *profit function* as $\pi(Q) = R(Q) - C(Q)$. Then, to maximize total profit, first set

$$\frac{dp}{dQ} = \frac{dR}{dQ} - \frac{dC}{dQ} = 0$$

Solutions Manual to Accompany Introduction to Quantitative Methods in Business: With Applications Using Microsoft® Office Excel®, First Edition. Bharat Kolluri, Michael J. Panik, and Rao N. Singamsetti.
© 2017 John Wiley & Sons, Inc. Published 2017 by John Wiley & Sons, Inc.

or

$$\frac{dR}{dQ} = \frac{dC}{dQ}.$$
$$\text{(MR)} \quad \text{(MC)}$$

This determines an output level $Q = Q_e$. Next, check Q_e in the second derivative since we require, for a maximum, that $d^2\pi/dQ^2 < 0$ at Q_e. In sum, at Q_e, $d\pi/dQ = 0$ and $d^2\pi/dQ^2 < 0$. Note that

a. if $P = $ constant, then $\pi(Q) = R(Q) - C(Q) = PQ - C(Q)$
   and

$$\frac{d\pi}{dQ} = P - \frac{dC}{cQ} = 0 \text{ or } P = MC;$$

b. if $P \neq$ constant, then

$$\pi(Q) = R(Q) - C(Q) = P(Q)Q - C(Q),$$

where $P(Q)$ represents the (inverse) demand curve. Hence,

$$\frac{d\pi}{dQ} = \frac{d}{dQ}(P(Q)Q) - \frac{dC}{dQ} \text{ or } MR = MC.$$

To maximize total revenue $R(Q)$, first set $dR/dQ = 0$ and determine Q. Then check that, at this level of Q, $d^2R/dQ^2 < 0$.

3.3   MODELS OF COST MINIMIZATION: INVENTORY COST FUNCTIONS AND ECONOMIC ORDER QUANTITY (EOQ)

A firm's optimal inventory level or optimal EOQ is determined by minimizing the *total inventory cost function:*

$$Y(Q) = \text{annual holding cost} + \text{annual ordering cost}$$
$$= AHC(Q) + AOC(Q)$$
$$= \frac{Q}{2}C_h + \frac{D}{Q}C_0,$$

where

Q = order quantity
$Q/2$ = average inventory level
C = unit cost of the product
C_h = unit holding cost per annum ($C_h = KC$, $K = $ constant);
D = annual demand
C_0 = fixed cost per order

Additionally, we have

$T =$ inventory period (cycle time);

$N = D/Q =$ orders per year.

The EOQ is determined where $Y(Q)$ is at a minimum or where annual holding cost equals annual ordering cost, that is, setting $AHC(Q) = AOC(Q)$ yields

$$Q^* = \text{optimal EOQ} = +\sqrt{\frac{2DC_0}{C_h}}.$$

Once Q^* is determined,

average inventory level $= Q^*/2$;

number of orders per year $= N = D/Q^*$;

cycle time in days $= 360/N$;

reorder point $= R = L \times$ daily demand $= L \times (D/360)$, where L is the lead time;

annual holding cost $= AHC(Q^*)$;

annual ordering cost $= AOC(Q^*)$; and

annual total cost $= Y(Q^*)$.

3.3.2   Solution Using the Calculus Approach

To minimize total inventory cost $Y(Q) = \frac{Q}{2}C_h + \frac{D}{Q}C_0$,
set

$$\frac{dY}{dQ} = \frac{1}{2}C_h - \frac{D}{Q^2}C_0 = 0$$

to obtain the optimal EOQ, Q^*, as

$$Q^* = +\sqrt{\frac{2DC_0}{C_h}}.$$

At Q^*, we require that

$$\frac{d^2Y}{dQ^2} = \frac{2DC_0}{Q^3} > 0$$

for a minimum of $Y(Q)$.

3.4 CONSTRAINED OPTIMIZATION: LINEAR PROGRAMMING

3.4.1 Linear Programming: Maximization

A typical maximization problem has the form

$$\text{maximize } Z = 1.5\,X_1 + X_2 \ (\text{objective function}) \text{ s.t.}$$

$$\left.\begin{array}{l} (1) \quad X_1 + X_2 \le 5, \\ (2) \quad 2X_1 + X_2 \le 7, \end{array}\right\} \text{structural constraints,}$$

$$X_1, X_2 \ge 0 \quad (\text{nonnegativity conditions}).$$

The region in the X_1, X_2- plane, which satisfies all of the constraints simultaneously (both the structural constrains and nonnegativity conditions), is the *feasible region* or region of admissible solutions (Figure 3.1).

Let us use the *complete enumeration of extreme points method* to solve the above problem. Table 3.1 houses the details.

Clearly, point A is the optimal extreme point since it renders the largest value of Z.

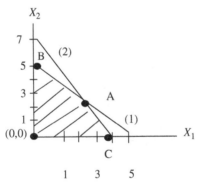

Figure 3.1 Maximizing Z.

Table 3.1 Objective Function Values

Extreme Point	$Z = 1.5\,X_1 + X_2$
(0,0)	$1.5\,(0) + 0 = 0$
A (2,3)	$1.5\,(2) + 3 = 6 = \max Z$
B (0,5)	$1.5\,(0) + 5 = 5$
C (3.5, 0)	$1.5\,(3.5) + 0 = 5.25$

Do either of the constraints display any excess capacity or slack at point A? To answer this question, let us calculate the values of the *slack variables*. That is, from the structural constraints,

$$\begin{rcl}(1) & X_1 + X_2 + X_3 = 5, \\ (2) & 2X_1 + X_2 + X_4 = 7,\end{rcl} \quad \text{or} \quad \begin{cases} X_3 = 5 - X_1 - X_2, \\ X_4 = 7 - 2X_1 - X_2. \end{cases}$$

At point A,

$$X_3 = 5 - 2 - 3 = 0,$$
$$X_4 = 7 - 2(2) - 3 = 0.$$

Hence, neither structural constraint displays any slack since the slack variables are zero. Both of these constraints are binding (hold as equalities) at the optimal solution.

3.4.2   Linear Programming: Minimization

A typical minimization problem has the form

$$\text{minimize } Z = 2X_1 + 2X_2 \text{ (objective function) s.t.}$$

$$\begin{rcl}(1) & 4X_1 + X_2 \geq 8, \\ (2) & X_1 + 3X_2 \geq 6,\end{rcl} \Big\} \text{ structural constraints,}$$

$$X_1, X_2 \geq 0 \quad \text{(nonnegativity conditions)}.$$

The region in the X_1, X_2-plane, which satisfies all of the constraints simultaneously, is termed the feasible region (Figure 3.2).

To solve this problem, let us employ the *complete enumeration of extreme points method*. Table 3.2 has the requisite calculations.

Obviously, point A represents the optimal extreme point since it yields the smallest value of Z.

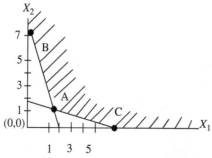

Figure 3.2   Minimizing Z.

Table 3.2   Objective Function Values

Extreme Point	$Z = 2\,X_1 + 2\,X_2$
A (1.65, 1.45)	$2(1.65) + 2(1.45) = 6.2 = \min Z$
B (0,8)	$2(0) + 2(8) = 16$
C (6, 0)	$2(6) + 2(0) = 12$

What about the values of the *surplus variables* at point A? To answer this question, let us specify the surplus variables in the following fashion. From the structural constraints,

$$\left.\begin{array}{l} (1) \quad 4X_1 + X_2 - X_3 = 8, \\ (2) \quad X_1 + 3X_2 - X_4 = 6, \end{array}\right\} \quad \text{or} \quad \left\{\begin{array}{l} X_3 = 4X_1 - X_2 - 8, \\ X_4 = X_1 + 3X_2 - 6. \end{array}\right.$$

At point A,

$$X_3 = 4(1.65) + 1.45 - 8 = 0,$$
$$X_4 = 1.65 + 3(1.45) - 6 = 0.$$

Hence, neither structural constraint displays any surplus (i.e., excess of the left-hand side over the right-hand side) since both surplus variables are zero. Thus, both structural constraints are binding (hold as equalities) at the optimal solution.

SOLUTIONS TO ODD-NUMBERED EXERCISES

1. Given the demand (price) and cost functions $P = 2000 - 40Q$ and $C(Q) = 3000 + 400Q$, respectively, find the following using the calculus approach:
   a. the profit maximizing output
   b. profit maximizing price
   c. maximum total profit
   d. the revenue maximizing output

SOLUTION:

1. It is obvious that total profit is at a maximum at $MR = MC$. So, let us try to find these functions and solve for the optimal level of Q.

$$R = PQ = (2000 - 40\,Q)Q$$
$$= 2000\,Q - 40\,Q^2.$$

So, $MR = (dR/dQ) = R'(Q) = 2000 - 40(2)Q^{2-1} = 2000 - 80Q.$
From the total cost function, $MC = dC/dQ = C'(Q) = 0 + 400Q^{1-1} = 400.$

a. Profit is at a maximum when $MR = MC$. It follows from this that

$2000 - 80Q = 400$

$80Q = 2000 - 400$

$80Q = 1600$

$Q = 20.$

Thus, we conclude that profit is maximized when $Q = 20$.

b. Then the corresponding price is

$$P = 2000 - 40\,(20)$$
$$= 2000 - 800$$
$$= 1200.$$

c. And maximum profit is

$$\text{Total revenue} - \text{total cost} = PQ - C(Q) = 1200 \times 20 - (3000 + 400(20))$$
$$= 24{,}000 - (3000 + 8000)$$
$$= 24{,}000 - 11{,}000$$
$$= 13{,}000.$$

d. Revenue maximizing output is obtained by setting $MR = 0$.
Thus, $MR = 2000 - 80Q = 0$.

$$80\,Q = 2000$$
$$Q \quad = \frac{2000}{80} = 25.$$

3. We noted earlier in this chapter that if a function $y = f(x)$ has a maximum or minimum at $x = x_0$, then $f'(x_0) = 0$. If $f' = 0$ is linear in x, then x_0 is readily determined by setting $f' = 0$ and solving for the value of x, which makes it vanish. But what if $f' = 0$ is nonlinear in x and cannot be easily factored? For example, what if it is, say, quadratic and of the form $0.2x^2 - 1.5x + 1 = 0$? To solve for the x values (*roots*) that satisfy this equation, let us employ the *quadratic formula*. To this end, suppose we have a general quadratic equation of the form

$$ax^2 + bx + c = 0, \quad a \neq 0.$$

Then it can be shown that

$$x = \frac{-b \pm \sqrt{b^2 - 4ac}}{2a}.$$

So, given $0.2x^2 - 1.5x + 1 = 0$ (here $a = 0.2$, $b = -1.5$, and $c = 1$),

$$x = \frac{1.5 \pm \sqrt{(1.5^2) - 4(0.2)(1)}}{2(0.2)} = \frac{1.5 \pm 1.204}{0.4}$$

so that $x = 6.76$ or $x = 0.74$.

Use the quadratic formula to find the roots of

a. $3x^2 + 5x - 7 = 0$

b. $-2x^2 + 3x + 5 = 0$

c. $3x^2 - x - 2 = 0$

SOLUTION:

3. **a.** $x = \dfrac{-5 \pm \sqrt{(5^2) - 4(3)(-7)}}{2(3)} = \dfrac{-5 \pm 10.440}{6}$

so that $x = -15.440$ or $x = 5.440$.

b. $x = \dfrac{-3 \pm \sqrt{(3^2) - 4(-2)(5)}}{2(-2)} = \dfrac{-3 \pm 7}{-4}$

so that $x = 2.5$ or $x = -1$.

c. $x = \dfrac{1 \pm \sqrt{((-1)^2) - 4(3)(-2)}}{2(3)} = \dfrac{1 \pm 5}{6}$

so that $x = -0.67$ or $x = +1$.

5. **a.** Given the total cost function $C(Q) = 0.01Q^3 - 3Q^2 + 4Q + 10$, determine the output level Q at which the marginal cost (MC) curve has a minimum. Does this output level correspond to the one for which the total cost function has a point of inflection? Verify your answer.

   b. Given $C(Q)$ above, find the average cost (AC) function. (*Note:* $AC(Q) = C(Q)/Q$.) At what Q value does average cost attain a minimum?

   c. Suppose the firm faces a constant price of $P = \$10$ for its product. Find the profit maximizing level of output. Now, let the firm face a demand (inverse) curve of the form $P = 10 - 3Q$ with the same total cost structure. What is the new profit-maximizing level of output? At what level of output is total revenue (TR) at a maximum?

SOLUTION:

5. **a.** $MC = C'(Q) = 0.03Q^2 - 6Q + 4$.

And to find a minimum of MC set $MC' = 0$.

$MC'(Q) = 0.06Q - 6 = 0$ or $Q = 6/0.06 = 100$. So, at $Q = 100$, MC achieves an extreme value. To see whether this value of Q corresponds to a maximum or a minimum, consider MC'' $(Q) = 0.06 \geq 0$ at any value of Q. So, MC is a minimum at $Q = 100$.

Since $C''(100) = MC'(100) = 0.06 \times 100 - 6 = 0$ and $C'''(100) = MC''(100) = 0.06 \neq 0$, the output level of $Q = 100$ is a point of inflection for the function $C(Q)$.

   b. The average cost function $AC(Q) = C(Q)/Q = (0.01Q^3 - 3Q^2 + 4Q + 10)/Q = 0.01Q^2 - 3Q + 4 + 10/Q$.

For AC to have a minimum, set $AC'(Q) = 0$ and solve for Q.

Now $AC'(Q) = 0.02Q - 3 - 10Q^{-2} = 0$. Multiplying this equation on both sides by Q^2, we get $0.02Q^3 - 3Q^2 - 10 = 0$. This is a cubic equation. One of the methods to solve this equation is by graphing it using Excel and finding the Q value at which $AC'(Q)$ crosses the Q-axis where $AC' = 0$.

You will see that the curve AC′ within the given range crosses the Q-axis around $Q = 150$, meaning that AC attains an extreme (minimum) value at $Q = 150$. That is, $AC'(150) = 0$. To see whether the sufficiency condition is satisfied, consider.

$AC''(Q) = 0.06Q^2 - 6Q$ and $AC''(150) = 0.06 \times 150^2 - 6 \times 150 = 450 > 0$, confirming that AC attains a minimum around $Q = 150$.

This is shown below.

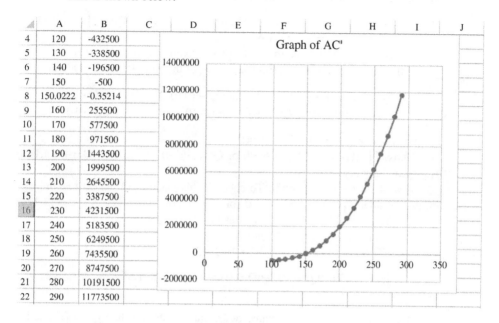

	A	B
4	120	-432500
5	130	-338500
6	140	-196500
7	150	-500
8	150.0222	-0.35214
9	160	255500
10	170	577500
11	180	971500
12	190	1443500
13	200	1999500
14	210	2645500
15	220	3387500
16	230	4231500
17	240	5183500
18	250	6249500
19	260	7435500
20	270	8747500
21	280	10191500
22	290	11773500

Instructions for drawing the above graph:
1. In column A enter Q values starting from 120 to say 290 with increments of 10.
2. In cell B4 enter formula = 0.02*highlightA4^3 − 3*highlightA4^2 − 10 and hit "Enter" key to get the value −432500 in B4 cell.
3. Go back to cell B4 and click and drag the fill handle (the right hand bottom corner when the cursor becomes black cross) to other cells in the B column.
4. Notice the values in column B change from negative to positive, thus attaining a zero value in between.
5. You can draw a graph by highlighting the data in both the columns A and B and click on "Insert" in the menu bar and select "Line Chart" to get the graph as shown above.

c. If the firm faces a constant price P = $10, then the profit function

$$\Pi(Q) = \text{Total revenue} - \text{Total cost}$$
$$= P \times Q - C(Q) = 10Q - (0.01Q^3 - 3Q^2 + 4Q + 10)$$
$$= -0.01Q^3 + 3Q^2 + 6Q - 10 \text{ is to be maximized.}$$

The first-order condition is $\Pi'(Q) = -0.03Q^2 + 6Q + 6 = 0$. Solving this quadratic equation we get

$$Q = \frac{-6 \pm \sqrt{(6)^2 - 4(-0.03)(6)}}{2(-0.03)} = \frac{-6 \pm 6.06}{-0.06}.$$

Thus, Q = 201 or Q = −1. The negative value of Q is ruled out since quantity sold is never negative.

The second-order condition for maximum profit is as follows:

$\Pi''\ (Q) = -0.06Q + 6$ should be negative at $Q = 201$. Thus, $\Pi''\ (201) = -0.06 \times 201 + 6 = -12.06 + 6$

$= -6.06 < 0$, confirming that the profit Π is maximized when $Q = 201$.

If the firm faces a demand (inverse) curve of the form $P = 10 - 3Q$, then

$$\Pi(Q) = P \times Q - C(Q) = (10 - 3Q)Q - (0.01Q^3 - 3Q^2 + 4Q + 10)$$
$$= 10Q - 3Q^2 - 0.01Q^3 + 3Q^2 - 4Q - 10 = -0.01Q^3 + 6Q - 10.$$

For profit maximization, set $\Pi'\ (Q) = 0$.

Thus, $\Pi'\ (Q) = -0.03Q^2 + 6 = 0$ or $Q^2 = -6/-0.03 = 200$ or $Q = \pm\ 14.14$. Ruling out any negative value,

$Q = 14.14$ maximizes profit. To check if this value of Q indeed maximizes profit, consider $\Pi''(Q) = -0.06Q < 0$ for any positive Q and so for $Q = 14.14$, thus confirming that this value maximizes profit.

At maximum TR (total revenue), TR' $(Q) = 0$.

Now $TR(Q) = PQ = (10 - 3Q)Q = 10Q - 3Q^2$ and

$$TR'(Q) = 10 - 6Q = 0 \text{ or } Q = 10/6 = 1.67.$$

The second-order condition for maximum TR is

TR'' $(Q) = -6 < 0$ for any value of Q, thus confirming that $Q = 1.67$ maximizes TR.

7. Suppose Q stands for order quantity and the annual demand for the product is 10,000 units. Assume that the cost of placing an order, independent of the size of the order, is $50. The cost of holding per unit is 50% of the unit cost of $2. Use this information to answer the following:

a. Show that the total annual cost of maintaining the inventory is

$$Y(Q) = \frac{Q}{2} + \frac{500,000}{Q}.$$

b. Find the economic order quantity, Q^*, and the total cost corresponding to that value of Q^*.

c. Also find the total cost if orders are placed for 1500 units.

d. Compare your results for (b) and (c) and comment.

SOLUTION:

7. a. Annual holding cost = Average inventory $\times$ C_h

$$= \frac{Q}{2} \times 1.$$

Annual ordering cost = number of orders per year $\times$ C_o

$$= \frac{D}{Q} \times 50$$

$$= \frac{10,000}{Q} \times 50 = \frac{500,000}{Q}.$$

Adding, the total inventory maintenance cost $= (Q/2) + (500,000/Q)$, which is to be established.

Here we have used the following information: $D = 10,000$, $C_h = 0.50 \times 2 = \$1$, and $C_0 = \$50$.

b. $\text{EOQ} = Q^* = \sqrt{\frac{2DC_0}{C_h}} = \sqrt{\frac{2 \times 10,000 \times 50}{1}} = 1000$.

Now from (a), annual cost of maintaining inventory is $(1000/2) + (500,000/1000) = \$500 + \$500 = \$1000$.

c. Now $Q = 1500$

$$\text{Annual holding cost} = \frac{Q}{2} \times C_h = \frac{1500}{2} \times 1 = \$750.$$

$$\text{Annual ordering cost} = \frac{D}{Q} \times C_o = \frac{10,000}{1500} \times 50 = 6.67 \times 50 = \$333.33.$$

Therefore, annual total cost = annual holding cost + annual ordering cost

$$= 750 + 333.33 = \$1083.33.$$

d. Comparing (b) and (c) we can see that we have an order quantity in part (c) different from EOQ of part (b), $Q^* = 1000$. The total inventory maintenance cost is $\$1083.33$, which is higher in part (c)

9. A firm estimates the annual demand for one of its products to be $12,000$ units. Each unit costs the firm $\$25.00$. The inventory holding cost per unit is 4% of the unit cost. The fixed cost of placing an order is $\$60.00$. The firm operates 360 days in the year. Currently the firm orders 1000 per month. Determine if the firm has to order more frequently or less frequently in order to minimize the annual inventory maintenance cost.

SOLUTION:

9. Based on the given information, the annual demand for the product is $D = 12,000$ units, the cost of holding per unit is $C_h = 0.04 \times 25 = 1$, and fixed cost of placing an order $C_o = 60$. Substituting these values into the EOQ formula, we obtain

$$Q^* = \sqrt{\frac{2DC_0}{C_h}} = \sqrt{\frac{2(12000)(60)}{1}}$$
$$= \sqrt{(12000)(120)} = \sqrt{1,440,000} = 1200.$$

Thus, the annual inventory maintenance cost is minimized at 1200 units. If the firm orders 1000 per month currently, it is not minimizing the total cost. The firm has to order less frequently at a level of 1200 units to minimize the cost.

11. Production to maximize profit:

Assume an electronics company in China is planning to produce two components used in the production of a major household appliance. Based on market experience, the estimated profits on the two components are given to be $\$11$ and $\$15$, respectively. The table below represents per unit labor and cash requirements and their availabilities.

Input Requirements and Availabilities

Product	Labor Hours	Cash ($)
Component 1	2	4
Component 2	3	5
Availability	12	22

Formulate as a linear programming model designed to maximize total profit. Also, present the graphical solution in detail. Identify slack variables in each of the structural constraints.

SOLUTION (by the complete enumeration of extreme points method):

11. The objective is to maximize total profit $Z = 11X_1 + 15X_2$, where X_1 the is number of units of component 1 and X_2 is the number of units of component 2 that are to be determined. The labor constraint is $2X_1 + 3X_2 \leq 12$ and the cash constraint is $4X_1 + 5X_2 \leq 22$. The mathematical model of the linear programming problem can be specified as follows.

$$\text{Maximize } Z = 11X_1 + 15X_2 \text{ s.t.}$$
$$\text{subject to } 2X_1 + 3X_2 \leq 12 \text{ (labor constraint)}$$
$$4X_1 + 5X_2 \leq 22 \text{ (cash constraint)}$$
$$X_1, X_2 \geq 0 \text{ (nonnegativity conditions).}$$

Step 1: Show the constraint inequalities graphically and identify the feasible region.

As required, the nonnegativity conditions restrict the solution sets of the structural constraints to the first quadrant of the graph. We may plot the linear constraint equations graphically by following the intercept method of Chapter 1. For example, consider the labor constraint equation $2X_1 + 3X_2 = 12$ and set $X_2 = 0$. This equation reduces to $2X_1 + 3(0) = 12$ and thus $X_1 = 6$, resulting in the point (6,0) in the following graph. Similarly, with $X_1 = 0$, we can solve for X_2 from $(0) + 3X_2 = 12$ so that $X_2 = 4$. Then, (0,4) is another point. Connecting these two points in the graph we obtain the labor constraint boundary line. Applying the same procedure to cash constraint, we find two points, (5.5, 0) and (0,4.4), on the cash constraint line and, by connecting these two points, we get the cash constraint boundary line. After drawing the straight lines corresponding to the boundaries of the constraints, we can easily identify the region representing the satisfaction of each of the constraints. We can now identify the region that constitutes the intersection of the sets of admissible solutions for these two "≤" structural constraints. This intersection obviously satisfies the direction of the inequalities and is shaded in the graph, which is also known as the feasible region.

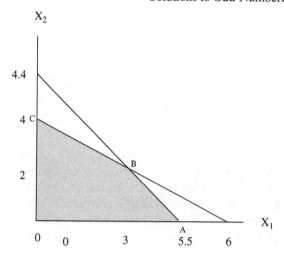

Step 2: Find the corner (extreme) points of the feasible region. The only unknown corner point is the point where the labor constraint and the cash constraint lines intersect. Obviously, it is obtained by solving the two constraint equations simultaneously.

To this end,

$$2\text{x(Labor constraint) yields } 4X_1 + 6X_2 = 24, \qquad (3.1)$$

$$\text{Cash constraint is } 4X_1 + 5X_2 = 22. \qquad (3.2)$$

Subtracting Equation (3.2) from the first we obtain $X_2 = 2$. Substituting this value into Equation (3.2) yields $4X_1 + 5(2) = 22$ or $4X_1 = 22 - 10 = 12$ and from this we find that $X_1 = 3$.

Therefore, the intersection point of the constraint lines has coordinates $(3, 2)$.

Step 3: Evaluate the objective function at each one of the extreme points and identify the optimum extreme point corresponding to the maximum value.

Extreme Point	$Z = 11X_1 + 15X_2$
O.$(0, 0)$	$11(0) + 15(0) = 0$
A.$(5.5, 0)$	$11(5.5) + 15(0) = 60.5$
B.$(3, 2)$	$11(3) + 15(2) = 63$
C.$(0, 4)$	$11(0) + 15(4) = 60$

It is clear from the above table that the company maximizes the total profit at $X_1 = 3$ units of component 1 and $X_2 = 2$ units of component 2, and the value of maximum total profit is 63. This is generally described as the optimal solution.

At the optimum point $(3, 2)$, the following table represents the values of the slack variables (X_3 and X_4) corresponding to the labor and cash constraints.

Slack Variables

Constraint	Left Side of the Constraint	Right Side of the Constraint	Slack Value
Labor	$2(3) + 3(2) = 12$	12	$X_3 = 0$
Cash	$4(3) + 5(2) = 22$	22	$X_4 = 0$

13. A dietician suggests a diet consisting of two food items to a particular patient in a hospital. The diet is designed to satisfy basic nutritional requirements at minimum cost. The two food items cost \$10.00/lb and \$12.00/lb, respectively. Food 1 contains 5% protein and 60% fat. Food 2 contains 30% protein and 40% fat. The patient needs a minimum of 2 lb of protein and at least 4.8 lb of fat per week. Formulate this as a linear programming model and find the food mix that minimizes the total cost of the diet.

SOLUTION (by the complete enumeration of extreme points method):

13. The objective is to minimize the cost of the diet, which is $10X_1 + 12X_2$, where X_1 is the number of pounds of food 1 and X_2 is the number of pounds of food 2. The constraints are $0.05X_1 + 0.3X_2 \geq 2$ lb for protein and $0.6X_1 + 0.4X_2 \geq 4.8$ lb for fat.
   Thus, the linear programming model is as follows:

$$\text{Minimize } Z = 10X_1 + 12X_2 \text{ s.t.}$$

$$0.05X_1 + 0.3X_2 \geq 2 \text{ (protein constraint)}$$

$$0.6X_1 + 0.4X_2 \geq 4.8 \text{ (fat constraint)}$$

$$X_1, X_2 \geq 0 \text{ (nonnegativity conditions)}.$$

Step 1: Show the constraint inequalities graphically and identify the feasible region.
   As required, the nonnegativity conditions restrict the solution sets of the structural constraints to the first quadrant of the graph. We may plot the linear constraint equations graphically by following the intercept method of Chapter 1. For example, consider the protein constraint equation $0.05 \ X_1 + 0.3 \ X_2 = 2$ and set $X_2 = 0$. This equation reduces to $0.05X_1 + 0.3 \ (0) = 2$ and thus $X_1 = 2/0.05 = 200/5 = 40$, resulting in the point (40,0). Similarly, with $X_1 = 0$, we can solve for X_2 from $0.05(0) + 0.3X_2 = 2$ so that $X_2 = 2/0.3 = 20/3 = 6.67$. Thus, (0,6.67) is another point. Connecting these two points in the figure below allows us to obtain the protein constraint boundary line. Applying the same procedure to the fat constraint, we find two points, (8, 0) and (0,12), on the fat constraint line and, by connecting these two points, we get the fat constraint boundary line. After drawing the straight lines corresponding to the boundaries of the constraints, we can easily identify the region representing the satisfaction of each of the constraints. It constitutes the intersection of the sets of admissible solutions for these two "$\geq$" structural constraints. This intersection obviously satisfies the direction of the inequalities and thus constitutes the (shaded) feasible region given in the figure below.

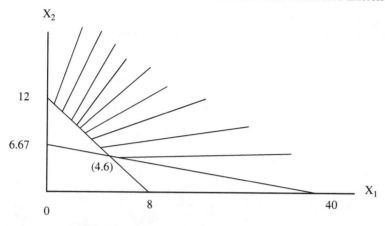

Step 2: Find the corner (extreme) points of the feasible region. The only unknown corner point is the point where the protein constraint and the fat constraint lines intersect. Obviously, it is obtained by solving the two constraint equations simultaneously.

To this end,

$$0.06x \text{ (protein constraint) yields } 0.03X_1 + 0.18X_2 = 1.2 \tag{3.1}$$

$$0.05x \text{ (protein constraint) results in } 0.03X_1 + 0.02 = 0.24. \tag{3.2}$$

Subtracting Equation (3.2) from the first equation we obtain $0.16X_2 = 0.96$. Therefore, $X_2 = 6$. Substituting this value into Equation (3.1) yields $0.03X_1 + 0.18$ $(6) = 1.2$ or $0.03X_1 + 1.08 = 1.2$. Here $0.03X_1 = 1.2 - 1.08 = 0.12$ and thus $X_1 = 4$.

Therefore, the intersection point of the protein and fat constraints has coordinates $(4, 6)$.

Step 3: Evaluate the objective function at each one of the extreme points and identify the optimum extreme point corresponding to the minimum value.

Extreme Point	$Z = 10X_1 + 12X_2$
$(0, 0)$	$10(0) + 12(0) = 0$
$(40, 0)$	$10(40) + 12(0) = 400$
$(4, 6)$	$10(4) + 12(6) = 112$

From the above table we obtain the optimum solution as $X_1 = 4$, $X_2 = 6$, and the total minimum cost is \$112.

At the optimum point $(4,6)$, the following table represents the values of the surplus variables (X_3, X_4) of the constraints.

Surplus Variables

Constraint	Left Side of the Constraint	Right Side of the Constraint	Slack Value
Protein	$0.05(4) + 0.3(6) = 2$	2	$X_3 = 0$
Fat	$0.6(4) + 0.4(6) = 4.8$	4.8	$X_4 = 0$

EXCEL APPLICATIONS

15. Find solutions to Exercise 6 in the textbook using Excel.

SOLUTION:

15.

	A	B	C	D	E	F
1	Ans(a)	692.8203		=sqrt(2*40*4800/0.8)		
2	Ans(b)	6.928203			=4800/B1	
3	Ans(c)	51.96152	=360/B2			
4	Ans(d)	66.66667		=5*4800/360		
5	Ans(e)	277.1281		=(B1/2)*08		
6	Ans(f)	277.1281			=40*B2	
7	Ans(g)	554.2563	=B5+B6			
8						

Ans means Answer:

17. a. Given the following information, calculate the following:

    (i) The profit maximizing output Q^*

    (ii) The corresponding price

    (iii) Maximum profit

    (iv) The output Q that maximizes revenue. Use Excel and the trial and error method

        $P(Q) = 2000 - 40Q$ and $C(Q) = 3000 + 400Q$. Use the trial values $Q = 15$–25.

        (*Hint:* Using Excel, you can find the solution values by trial and error. Refer to the examples in the textbook, Examples 3.1 and 3.2, and develop a similar table.)

SOLUTION:

17. a. Computation of Profit, MR, and MC using Excel

    From the above table we can see that

    (i) Profit maximizing output is $Q = 20$, where $MR = MC$.

    (ii) Profit maximizing price is $P = 1200$, when $Q = 20$.

    (iii) Maximum total profit is $TP = TR - TC = 13,000$.

    (iv) Revenue maximizing output $Q = 25$ (approximately), where $MR = 0$.

b. Repeat Exercise 17a with the following functions: $P = 300 - Q$ and $C(Q) = 4000 + 100Q$ in the range of $Q = 95$–105 with increments of 1 unit. Present the table showing detailed computations of profit, MR, and MC using Excel.

| =2000-40*A2 | =A2*B2 | =3000+400*A2 | =C2-D2 | =(C3-C2)/(A3-A2) | =(D3-D2)/(A3-A2) |

	A	B	C	D	E	F	G	H	I	J
1	Q	P	TR	TC	Totalprofit	MR	MC	Tracking	Criterion	
2	15	1400	21000	9000	12000			NA	NA	
3	16	1360	21760	9400	12360	760	400	MR>MC	MR-MC > 0	
4	17	1320	22440	9800	12640	680	400	MR>MC	MR-MC > 0	
5	18	1280	23040	10200	12840	600	400	MR>MC	MR-MC > 0	
6	19	1240	23560	10600	12960	520	400	MR>MC	MR-MC > 0	
7	20	1200	24000	11000	13000	440	400	MR>MC	MR-MC > 0	
8	20.01	1199.6	24004	11004	13000	400	400	MR=MC	MR-MC = 0	
9	21	1160	24360	11400	12960	360	400	MR<MC	MR-MC < 0	
10	22	1120	24640	11800	12840	280	400	MR<MC	MR-MC < 0	
11	23	1080	24840	12200	12640	200	400	MR<MC	MR-MC < 0	
12	24	1040	24960	12600	12360	120	400	MR<MC	MR-MC < 0	
13	25	1000	25000	13000	12000	40	400	MR<MC	MR-MC < 0	
14	25.01	999.6	25000	13004	11996	0				
15										
16										
17										
18										
19										
20										
21										

SOLUTION:

17.b

Q	P	TR	TC	TP	MR	MC	Tracking	Criterion
95	205	19,475	13,500	5975				
96	204	19,584	13,600	5984	109	100	MR > MC	MR − MC > 0
97	203	19,691	13,700	5991	107	100	MR > MC	MR − MC > 0
98	202	19,796	13,800	5996	105	100	MR > MC	MR − MC > 0
99	201	19,899	13,900	5999	103	100	MR > MC	MR − MC > 0
100	200	20,000	14,000	6000	101	100	MR > MC	MR − MC > 0
100.000001	200	20,000	14,000	6000	100	100	MR = MC	MR − MC = 0
101	199	20,099	14,100	5999	99	100	MR < MC	MR − MC < 0
102	198	20,196	14,200	5996	97	100	MR < MC	MR − MC < 0
103	197	20,291	14,300	5991	95	100	MR < MC	MR − MC < 0
104	196	20,384	14,400	5984	93	100	MR < MC	MR − MC < 0
105	195	20,475	14,500	5975	91	100	MR < MC	MR − MC < 0

From the above table we can see that
  (i) Profit maximizing output Q = 100, where MR = MC.
  (ii) Profit maximizing price is P = 200, when Q = 100.
  (iii) Maximum total profit TP = TR − TC = 6000.
  (iv) Revenue maximizing output is Q = 105 (approximately), where MR = 0.

19. a. A hospital dietician formulates an eating plan consisting of three food items. The plan purports to satisfy the requisite nutritional requirements at minimum cost. The three food items cost $10.00/lb, $12.00/lb, and 16/lb, respectively. Food 1 contains 5% protein and 60% fat. Food 2 contains 30% protein and 40% fat. Food 3 contains 40% protein and 30% fat. The patient needs a minimum of 2 lb of protein and at least 4.8 lb of fat per week. Formulate this as a linear programming model and find the food mix that minimizes the total cost of the diet using the Solver Program in Excel.

b. In 19(a) above, suppose that due to contractual obligations, the hospital has to order at least 1 lb of food 3 per week. Find the optimal solution under this additional constraint.

SOLUTION:

19. a. The objective function is to minimize the cost of the die, which is $10X_1 + 12X_2 + 16X_3$, where X_1 is the number of pounds of food 1, X_2 is the number of pounds of food 2, and X3 is the number of pounds of food 3. The constraints are

$$0.05X_1 + 0.3X_2 + 0.4X_3 \geq 2 \text{ lb for protein}$$

and

$$0.6X_1 + 0.4X_2 + 0.3X_3 \geq 4.8 \text{ lb for fat.}$$

Now the complete mathematical model is as follows:
Minimize $Z = 10X_1 + 12X_2 + 16X_3$ s.t.
$0.05X_1 + 0.3X_2 + 0.4X_3 \geq 2$ lb (protein constraint)
$0.6 X_1 + 0.4X_2 + 0.3X_3 \geq 4.8$ lb (fat constraint)
$X_1, X_2, X_3 \geq 0$ (nonnegativity conditions).

Step 1: Open a new Excel spreadsheet.
Step 2: Go to Excel Options and open the Add-Ins tab.
Step 3: Find the Solver Add-in and click Go.

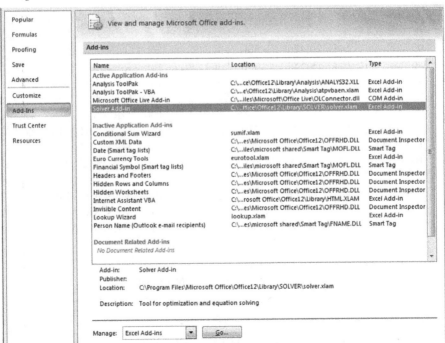

Step 4: Find the Solver Add-in box, check it and click OK as below.

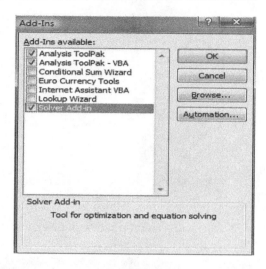

Step 5: Set up the spreadsheet as below and click in blank cell before going to step 6.

	E3			f_x =SUMPRODUCT($B$2:$D$2,B3:D3)		
	A	B	C	D	E	F	G
1		x1	x2	x3			
2	decision	1	1	1			
3	obj. function	10	12	16	38		
4	protein constraint	0.05	0.3	0.4	0.75	>=	2
5	fat constraint	0.6	0.4	0.3	1.3	>=	4.8
6							

Step 6: Go to solver and open it. Plug in the parameters as below:

Step 7: Go to options and click Assume Linear Model and Assume Nonnegative.

Step 8: Click on "Solve" in the Solver parameters dialog box to solve.

	E3		f_x =SUMPRODUCT($B$2:$D$2,B3:D3)				
	A	B	C	D	E	F	G
1		x1	x2	x3			
2	decision	4	6	0			
3	obj. function	10	12	16	112		
4	protein constraint	0.05	0.3	0.4	2 >=		2
5	fat constraint	0.6	0.4	0.3	4.8 >=		4.8
6							

Optimal decision is to buy 4 lb of food 1, 6 lb of food 2, and 0 lb of food 3 at a minimum cost of $112 for nutritional requirements to be satisfied.

b. Suppose that due to contractual obligations the hospital has to order at least 1 lb of food 3 per week. Find the optimal solution under this additional constraint.

$$X_3 \geq 1 \text{ (contractual obligation).}$$

Step 1: Go to the spreadsheet that you prepared for (a). Add the additional constraint to your spreadsheet and click on an empty cell before going to the next step.

	E3		f_x =SUMPRODUCT($B$2:$D$2,B3:D3)				
	A	B	C	D	E	F	G
1		x1	x2	x3			
2	decision	1	1	1			
3	obj. function	10	12	16	38		
4	protein constraint	0.05	0.3	0.4	0.75 >=		2
5	fat constraint	0.6	0.4	0.3	1.3 >=		4.8
6	contectual obligation	0	0	1	1 >=		1
7							

Step 2: Go to solver and plug in the parameters.

Step 3: Go to options and click Assume Linear Model and Assume Nonnegative.

Step 4: Click "Solve" button in step 2 above to get the following result.

	E3		f_x	=SUMPRODUCT($B$2:$D$2,B3:D3)		
	A	B	C	D	E	F	G
1		x1	x2	x3			
2	decision	4.4375	4.59375	1			
3	obj. function	10	12	16	115.5		
4	protein constraint	0.05	0.3	0.4	2 >=		2
5	fat constraint	0.6	0.4	0.3	4.8 >=		4.8
6	contectual obligation	0	0	1	1 >=		1

Optimal solution is $X_1 = 4.4375$ lb, $X_2 = 4.59375$ lb, and $X_3 = 1$ lb. Minimum cost is $115.5.

Chapter 4

What Is Business Statistics?

4.3 DESCRIPTIVE STATISTICS: TABULAR AND GRAPHICAL TECHNIQUES

A basic method for summarizing a set of data is to form a *frequency table*. It shows the number of observations in a given data set that fall into each of several mutually exclusive and exhaustive classes. How is a frequency table constructed? We shall follow a three-step procedure:

Step 1: Determine the number of classes. A guide for completing this step is: if an integer k is the number of classes and n is the size of the data set, then k is determined as the smallest integer such that $2^k \geq n$.

Step 2: Determine the width of each class or the *class interval*. This is accomplished by dividing the range of the data set by the number of classes (from step 1) and rounding off to a higher number. Typically, we work with an equal width for each class.

Step 3: Form the frequency table. Begin by selecting the *lower limit* of the first class as a number equal to or smaller than the minimum value in the data set. Add to this number the (common) class length to obtain the lower limit of the second class. This process is continued until all the lower class limits are set. Since the classes must not overlap, the *upper class limits* are chosen accordingly.

Once the classes are specified, we count the number of data values falling into each class so as to determine the *class frequencies*. The resulting construction is known as a *frequency distribution* for grouped data.

A *relative frequency distribution* is obtained by using the same classes but with the (absolute) class frequencies replaced by the relative class frequencies, that is, if f_j is the frequency of the jth class, then the *relative frequency* of that class is f_j/n.

Solutions Manual to Accompany Introduction to Quantitative Methods in Business: With Applications Using Microsoft® Office Excel®, First Edition. Bharat Kolluri, Michael J. Panik, and Rao N. Singamsetti.

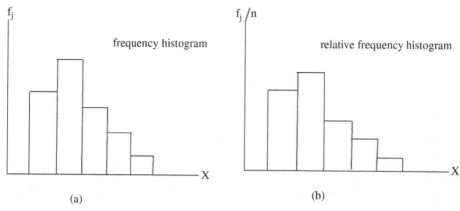

Figure 4.1   Histograms.

Both the frequency distribution and the relative frequency distribution can be illustrated graphically by a *histogram* (Figure 4.1a and b, respectively). Here the class intervals are shown on the x-axis, where it is to be understood that the *class boundaries* are used to replace the actual class limits in order to avoid any gaps between the classes.

An alternative graphical technique for illustrating a frequency or relative frequency distribution is a *frequency polygon* or a *relative frequency polygon* (Figure 4.2 a and b, respectively). Here, the *midpoints* (m_j) of the class intervals (m_j is the average of the class limits of class j) are plotted on the x-axis.

A *cumulative frequency distribution* is obtained by using the chosen classes but with the actual class frequencies replaced by a "running total" of the class frequencies. Its graphical description is called a *cumulative frequency curve* (OGIVE) and is obtained by plotting the upper boundaries of the classes against the corresponding cumulative frequencies (Figure 4.3).

A *cumulative relative frequency curve* is defined in an analogous fashion.

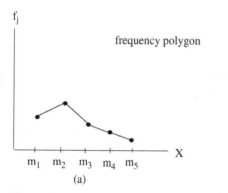

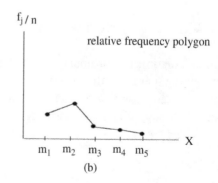

Figure 4.2   Frequency polygons.

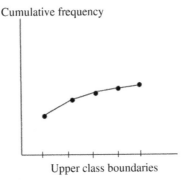

Figure 4.3   Cumulative frequency curve.

4.4 DESCRIPTIVE STATISTICS: NUMERICAL MEASURES OF CENTRAL TENDENCY OR LOCATION OF DATA

4.4.1 Population Mean

We previously defined a *population* as the whole set of objects of interest, that is, the collection of all observations on some variable X. If we denote the population size as N, then the *population mean* is calculated as

$$\mu = \frac{\sum X_i}{N}.$$

4.4.2 Sample Mean

We know that a *sample* is a subset of the population. For a sample of size $n < N$, the sample mean is calculated as

$$\overline{X} = \frac{\sum X_i}{n}.$$

4.4.3 Weighted Mean

In calculating the ordinary mean of a sample or population, we are implicitly assuming that all of the observations have the same relative importance. But what if some data points are deemed more important than others? To take account of the differing relative importance of the observations, we need to calculate a *weighted mean*

$$\overline{X}_w = \frac{\sum X_i W_i}{\sum W_i},$$

where the weights W_i are chosen to represent relative importance. (*Note:* If all of the weights are equal or $W_i = 1/n$ for all i, then $\overline{X}_w = \overline{X}$.)

4.4.4 Mean of a Frequency Distribution: Grouped Data

The *mean of a frequency distribution for grouped data* is calculated as

$$\overline{X} = \frac{\sum_{j=1}^{K} f_j m_j}{\sum_{j=1}^{K} f_j} = \frac{\sum_{j=1}^{K} f_j m_j}{n},$$

where f_j is the frequency of the jth class and m_j is the *class mark* or *midpoint* of the jth class.

4.4.5 Geometric Mean

For a set of positive data values $X_1, X_2, \ldots, X_n$, the nth root of the product of these n values is termed the *geometric mean* and is calculated as

$$\overline{X}_g = \sqrt[n]{X_1 \cdot X_2 \cdots X_n}.$$

When dealing with financial assets that increase in value each year at different rates, the geometric mean formula can be written as

$$\overline{X}_R = \sqrt[n]{(1 + R_1)(1 + R_2) \cdots (1 + R_n)} - 1,$$

where R_i is the growth rate in year i and n is the number of periods.

4.4.6 Median

The *median* is the middle value of a data set once the data points have been arranged in an increasing sequence. So, if $X_1, X_2, \ldots, X_n$ are the *ordered* data values, then

a. for an *odd* number of observations, there is always a middle point whose value is the median. Its position is located as

$$X_{(n+1)/2}.$$

b. for an even number of observations, the median is located at the position

$$\left(X_{(n/2)} + X_{(n/2)+1} \right)/2.$$

4.4.7 Quantiles, Quartiles, Deciles, and Percentiles

When an *ordered* data set is partitioned into four equal parts, the three points of division are called *quartiles* and denoted $Q_1, Q_2,$ and Q_3, that is, the first quartile

Q_1 indicates that 25% of the observations lie below Q_1, and 75% of the data values are equal to or lie above Q_1. Q_2 and Q_3 are interpreted in a similar fashion. Note that the second quartile Q_2 coincides with the median of a data set. Similarly, 9 deciles $(D_1, D_2, \ldots, D_9)$ split the set of observations into 10 equal parts, for example, the first decile D_1 indicates that 10% of the data values lie below D_1, and 90% of these data points are equal to or lie above D_1, and so on.

Percentiles, which are 99 in number (they are denoted as $(P_1, P_2, \ldots, P_{99})$, split the *ordered* data values into 100 equal segments. The ith percentile is located at the position

$$P_i = \frac{i(n+1)}{100}, \quad i = 1, \ldots, 99,$$

where n is the number of data points. For instance, the first percentile P_1 reveals that 1% of the data points lie below P_1, and 99% of the observations are equal to or lie above P_1.

It is important to remember that percentiles, deciles, and quartiles are *positional and not computed values*. Additionally, there is some commonality between these *quantiles*. For instance,

$$Q_1 = P_{25},$$

$$Q_2 = P_{50} = D_5 = \text{median},$$

$$Q_3 = P_{75}, \quad \text{and}$$

$$D_1 = P_{10}, \quad D_2 = P_{20}, \quad \ldots, \quad D_9 = P_{90}.$$

4.4.8 Mode

The *mode* is defined as the most common or most frequently occurring value in a data set. The mode may not exist, but if it does exist, it may not be unique, and, unlike the mean and median, it *always* corresponds to one of the actual data points (provided it exists).

Comparison of the Mean, Median, and Mode

1. The mean = median = mode if the distribution of the data set is mound-shaped and symmetric (Figure 4.4a).
2. A distribution is said to be *skewed* if it departs from symmetry. Given that the distribution is still mound-shaped,
   a. if mean > median > mode, the data set is termed *right skewed* or *positively skewed* (Figure 4.4b).
   b. if mean < median < mode, the data set is termed *left skewed* or *negatively skewed* (Figure 4.4c).

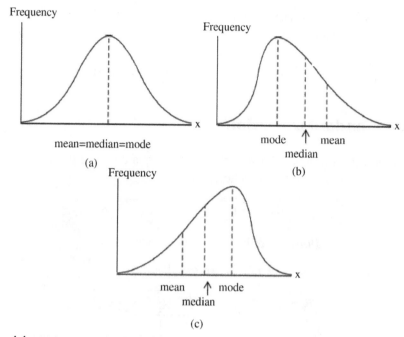

Figure 4.4   (a) Symmetric and mound-shaped distribution. (b) Right-skewed distribution. (c) Left-skewed distribution.

4.5   DESCRIPTIVE STATISTICS: MEASURES OF DISPERSION (VARIABILITY OR SPREAD)

4.5.2   Variance

The *variance* is the average of the squared deviations of the observations from the mean. For a set of population values $X_1, X_2, \ldots, X_N$, the *population variance* is specified as

$$\sigma^2 = \sum_{i=1}^{N} (X_i - \mu)^2 / N \quad \text{(long formula)},$$

where $X_i - \mu$ is the ith deviation from the population mean $\mu = \sum_{i=1}^{N} X_i / N$. Alternatively, σ^2 may be written as

$$\sigma^2 = \frac{1}{N} \sum_{i=1}^{N} X_i^2 - \mu^2 \quad \text{(short formula)}.$$

For a sample of observations $X_1, X_2, \ldots, X_n$, the *sample variance* appears as

$$S^2 = \frac{\sum_{i=1}^{n} (X_i - \overline{X})^2}{n - 1} \quad \text{(long formula)},$$

where $X_i - \overline{X}$ is the ith deviation from the sample mean $\overline{X} = \sum_{i=1}^{n} X_i/n$. An alternative representation for S^2 is

$$S^2 = \frac{\Sigma X_i^2 - [(\Sigma X_i)^2/n]}{n-1} \quad \text{(short formula)}.$$

4.5.3 Standard Deviation

The *standard deviation* is defined as the positive square root of the variance. The *population standard deviation* is denoted as

$$\sigma = \underbrace{\sqrt{\frac{\Sigma(X_i - \mu)^2}{N}}}_{\text{(long formula)}} = \underbrace{\sqrt{\frac{1}{N}\Sigma X_i^2 - \mu^2}}_{\text{(short formula)}}.$$

And the *sample standard deviation* is expressed as

$$S = \underbrace{\sqrt{\frac{\sum(X_i - \overline{X})^2}{n-1}}}_{\text{(long formula)}} = \underbrace{\sqrt{\frac{\sum X_i^2}{n-1} - \frac{\sum(X_i)^2}{n(n-1)}}}_{\text{(short formula)}}.$$

At times it may be necessary to determine the standard deviation of a sample of observations presented in *grouped form*, that is, we have a distribution involving classes of observations and their corresponding frequencies. We previously noted that the sample mean for this type of data set appears as

$$\overline{X} = \frac{\sum_{j=1}^{K} f_j m_j}{\sum_{j=1}^{K} f_j},$$

where m_j is the class mark or midpoint of the jth class and f_j is the frequency of the jth class, $j = 1, \ldots, K$. Then,

$$S = \sqrt{\frac{\sum_{j=1}^{K} f_j (m_j - \overline{X})^2}{n-1}} \quad \text{(long formula)}$$

and

$$S = \sqrt{\frac{\sum_{j=1}^{K} f_j m^2 - \left[\left(\sum_j^{K} f_j m_j\right)^2/n\right]}{n-1}} \quad \text{(short formula)}.$$

4.5.4 Coefficient of Variation

The standard deviation is a measure of *absolute variation*. In this regard, for two or more data sets, we cannot compare their variability if these data sets have different means and/or are expressed in different units. This is why we use a measure of *relative variation* called the *coefficient of variation*. It is defined as

$$CV = \left(\frac{\text{standard deviation}}{\text{mean}}\right) \times 100\%$$

given that the mean is positive. (If the mean happens to be negative, simply ignore the sign of CV.) For a

$$\text{Population-CV} = \frac{\sigma}{\mu} \times 100\%,$$

$$\text{Sample-CV} = \frac{S}{\overline{X}} \times 100\%.$$

How is the CV interpreted? Suppose we have a sample and that $CV = 22\%$. Then, we conclude that the standard deviation is 22% of the mean.

4.5.5   Some Important Uses of the Standard Deviation

1. *Standardization of values*

   To *standardize* a particular data point X_i, we form its Z-score, that is,

   $$\text{Population-}Z_i = \frac{X_i - \mu}{\sigma},$$

   $$\text{Sample-}Z_i = \frac{X_i - \overline{X}}{S}.$$

   Here Z_i represents the distance of X_i from the mean in terms of standard deviation units. If $Z_i > 0$, then X_i lies above the mean; if $Z_i < 0$, X_i lies below the mean. For instance, if $Z_i = 2.5$, then X_i lies two and one half standard deviations *above* the mean. And if $Z_i = -1$, then X_i lies one standard deviation *below* the mean.

2. *Chebysheff's theorem*

   A useful concept for describing the distribution of a data set is Chebysheff's theorem. For *any* set of data and *any* constant $K > 1$, at least $1 - (1/K^2)$ of the observations must lie within K standard deviations of the mean. The implied interval (for a sample of data) for $K > 1$ is $\overline{X} \pm KS$ or $(\overline{X} - KS, \overline{X} + KS)$. Clearly, this interval is symmetrical about the mean.

4.5.6   Empirical Rule

The empirical rule that follows applies to a mound-shaped data set. Specifically, for a sample,

1. approximately 68% of the observations will fall within one standard deviation of the mean or within the interval $(\overline{X} - S, \overline{X} + S)$;

2. about 95% of the observations will fall within two standard deviations of the mean or within the interval $(\overline{X} - 2S, \overline{X} + 2S)$; and

3. about 99.7% of the data points will fall within three standard deviations of the mean or within the interval $(\overline{X} - 3S, \overline{X} + 3S)$.

Note: While this rule applies *only* to mound-shaped data sets, Chebysheff's theorem applies to *any* data set or distribution. Moreover, the empirical rule is actually a special case of Chebysheff's theorem.

4.6 MEASURING SKEWNESS

The concept of *skewness* relates to the *shape* of a distribution, that is, if a distribution is skewed, then it lacks symmetry. (*Note:* A distribution is *symmetrical* if, when "folded" at the median, the left-hand side is the mirror image of the right-hand side.)

To measure the degree of skewness, we can use the *coefficient of skewness*:

$$SK = \frac{\text{mean} - \text{mode}}{S}.$$

If the mode cannot be uniquely determined, use

$$SK = \frac{3(\text{mean} - \text{median})}{S}.$$

SOLUTIONS TO ODD-NUMBERED EXERCISES

1. The following data set represents merchandise trade as a percentage of GDP of 100 countries in year 2014. *Data file:* Merchandise.xlsx

Table 4.1

40	57	25	54	33	51	102	42	101	173
75	66	92	20	81	115	50	30	41	145
34	53	55	58	42	31	49	67	37	159
61	43	53	33	56	45	70	57	71	70
42	44	50	53	32	90	51	99	157	61
38	40	34	67	77	47	47	33	87	56
40	78	100	56	80	97	51	102	136	77
50	63	96	92	61	65	102	91	44	145
103	46	28	46	29	74	74	40	45	79
62	81	43	69	61	49	41	111	57	40

Source: http://data.worldbank.org/2015.

(Merchandise Trade as a share of GDP is the sum of merchandise exports and imports divided by the value of GDP, all in current U.S. dollars.)

For the above data obtain the following:

a. Frequency distribution

b. Relative and cumulative frequency distributions

c. Histograms

d. Frequency polygon

e. Ogive curve

SOLUTION:

a.

Step 1: Determine the number of classes (groups).

The number of classes chosen is arbitrary, but should not be too few or too many. Usually, the number of classes chosen is anywhere between 5 and 15, depending on the size of the data set. For example, if we have a data set of 20 values, then we can arrange them into, say, five classes. Likewise, if we have a data set of $200,000$ values, we may select a larger number of classes such as15. There is also a mathematical formula that can be used as a guide for determining the number of classes. If k is the number of classes (an integer) and n is the number of values in the data set, then k is determined as the smallest integer satisfying $> 2^k \geq n$. For example, if $n = 35$, then k is 6, as is evident from the following table:

Table 4.2   Sample Size and the Number of Classes

n	k	2^k
10	4	16
20	5	32
35	6	64
70	7	128

In our example, the data size is n $= 100$ and therefore k $= 7$ since $2n > 100$.

Step 2: Determine the class width or class interval.

The class interval or length of a class is obtained by dividing the range of the data set by the number of classes obtained in step 1 above and rounding up to a higher and convenient number. Following this procedure, we determine the range as the difference between the largest and the smallest values in the data set. In our example, range $= 173 - 20 = 153$. Dividing this by the number of classes, $153/7$, we obtain 21.9, or approximately 22. Note that this approximation of class width is always toward a higher number in order to accommodate all the values in the data set. Also, note that we are assuming an equal width for all classes. It is possible to have examples with unequal width as well as with the first and/or last classes being open ended. The latter case can arise when there are no particular limits for low or high values, for example, in an age distribution, one of the classes might be "65 and over."

Step 3: Form the frequency, relative frequency, and cumulative frequency tables.

We begin by selecting the lower limit of the first class as a convenient number equal to or smaller than the minimum value of the data set. Thus, in our case, since the minimum is 20, we select 19 as the lower limit of the first class. (Note that the lower limit of a class is the smallest value going into the class.) Since the length of each class is 22, the lower limit of the second class must be $19 + 22 = 41$. Continuing in this fashion we see that the lower limit of the third class is $41 + 22 = 63$; the lower limit of the fourth class is $63 + 22 = 85$; and so on. Since the classes must not overlap (each observation belongs to one and only one class), the upper limit of the first class must be 40. Thus, the upper limit of the second class is $62 = 40 + 22$, and the upper limit of the third class is $62 + 22 = 84$, and so on. (Clearly, the upper limit of a class is the largest value that the class contains.) This procedure yields the following seven classes, where the class from 151–173 accommodates the largest value in the data set:

$$19 - 40,$$
$$41 - 62,$$
$$63 - 84,$$
$$85 - 106,$$
$$107 - 128,$$
$$129 - 150, \text{ and}$$
$$151 - 173.$$

Once the classes are specified, we sort each of the 100 percentage values of the merchandise into the classes, and then count the actual number of items in each class so as to obtain the class frequencies, as shown in the following table. The resulting construction is known as a frequency distribution or frequency table for grouped data and consists of columns two and four. The other columns are either work columns or represent ancillary calculations.

Table 4.3a   Frequency Distribution

Serial No.	Class	Tally	Frequency
1	19–40	卌 卌 卌 |||	18
2	41–62	卌 卌 卌 卌 卌 卌 卌 卌	40
3	63–84	卌 卌 卌 ||||	19
4	85–106	卌 卌 ||||	14
5	107–128	||	2
6	129–150	||||	4
7	151–173	|||	3
Total			100

b. Relative and cumulative frequency distributions

The relative frequency column is obtained by dividing the frequency in each class by the total number of values, n = 100; and the cumulative frequency column is obtained as shown in the following table:

Table 4.3b Relative Frequency and Cumulative Frequency Table

Serial No.	Class	Frequency	Relative Frequency	Cumulative Frequency
1	19–40	18	.18	18
2	41–62	40	.40	$18+40=58$
3	63–84	19	.19	$58+19=77$
4	85–106	14	.14	$77+14=91$
5	107–128	2	.02	$91+2=93$
6	129–150	4	.04	$93+4=97$
7	151–173	3	.03	$97+3=100$
Total		100	1.00	

c. Histograms

The frequency distribution can also be shown in graphical form. This is accomplished by showing class intervals on the x-axis and the frequencies on the y-axis, with appropriate scales. It is to be noted that the class limits are sometimes replaced by class boundaries in order to make the histogram continuous. Class boundaries do not exist in the actual data. They are utilized to avoid gaps between the classes. (An alternative way to eliminate gaps between the classes is to exclude the upper limits in each class and to include only the lower limits. For example, in the first table above, the first class could be 19–41, the second one, 41–63, etc.) For this frequency distribution, the class boundaries are as follows:

18.5–40.5
40.5–62.5
62.5–84.5
84.5–106.5
106.5–128.5
128.5–150.5
150.5–173.5

(Note that the length of a class can be obtained by taking the difference between its upper and lower boundaries.) The following figure shows the frequency histogram for our example, where it is implicitly understood that the base of each rectangle goes from its lower boundary to its upper boundary. This eliminates any gaps between the rectangles.

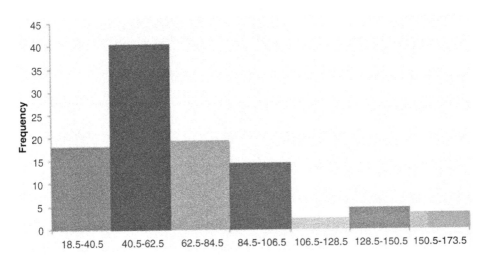

Figure 4.1 Merchandise trade

A similar histogram, known as a relative frequency histogram (probability chart), can also be drawn using the class intervals on the x-axis and the relative frequencies shown in the second table above on the y-axis. This is shown in the following graph:

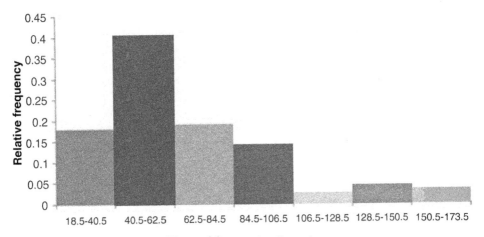

Figure 4.2 Merchandise trade

d. Frequency polygon

A frequency polygon is obtained by plotting the midpoints of the class intervals (the average of the class limits) against their frequencies and connecting them by straight lines, as shown in the following graph:

Table 4.3c Frequency Table with Midpoints

Serial No.	Class	Midpoint	Frequency
1	19–40	29.5	18

2	41–62	51.5	40
3	63–84	73.5	19
4	85–106	95.5	14
5	107–128	117.5	2
6	129–150	139.5	4
7	151–173	162	3
Total			100

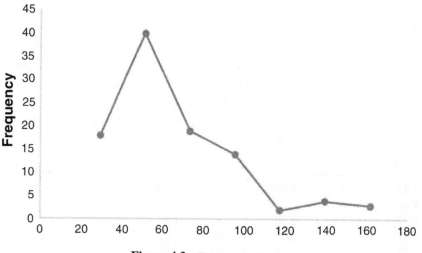

Figure 4.3   Frequency polygon

e. Ogive

The cumulative frequency curve (ogive) is obtained by plotting the upper boundaries of each class against the corresponding cumulative frequency. This curve is useful in answering questions such as "How many countries are there at or below a certain percentage merchandise trade level of GDP?" For instance, in this example, there are 91 countries whose merchandise trade is at or below 106.5%. The cumulative frequencies and the cumulative frequency curve are shown in the following table.

Table 4.4   Cumulative Frequency Values

Upper Class Boundaries (%)	Cumulative Frequencies (Less Than or Equal to)
40.5	18
62.5	58
84.5	77
106.5	91
128.5	93
150.5	97
173.5	100

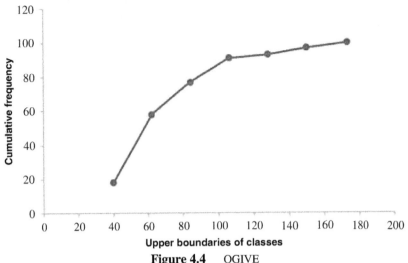

Figure 4.4   OGIVE

3. The following data set represents new home mortgage yields in the United States from 1963 to 2014. *Data file:* Mrtgyld.xlsx

Table 4.7

5.89	5.83	5.81	6.25	6.46	6.97	7.81	8.45
7.74	7.60	7.96	8.92	9.0	9.0	9.02	9.56
10.78	12.66	14.70	15.14	12.57	12.38	11.55	10.17
9.31	9.19	10.13	10.05	9.32	8.24	7.20	7.49
7.87	7.8	7.71	7.07	7.04	7.52	7.0	6.43
5.80	5.77	5.94	6.63	6.41	6.05	5.14	4.80
4.56	3.69	4.00	4.22				

Source: Economic Report of the President, 2015, Table B-17.

For the above data, obtain the following:

a. Frequency distribution

b. Relative and cumulative frequency distributions

c. Histograms

d. Frequency polygon

e. Ogive

SOLUTION:

a. Frequency distribution

To develop the frequency distribution, we follow the three steps described in Example 4.1.

Step 1: Determine the required number of classes (groups) k.

By using the formula $2^k \geq n$, we find that the number of classes, k, is 6 since

$n = 52$ (sample size), $2^6 = 64$, and thus $2^6 \geq 52$.

Step 2: Determine the class width or class interval.

It is obtained by dividing the range of the data set by the number of classes and rounding to a higher convenient number. We know that the range is the difference between the highest and the lowest values in the data set, or $15.14 - 3.69 = 11.45$. Dividing this difference by the number of classes, we obtain $11.45/6 = 1.91$ rounded-off to 2.

Step 3: Form the frequency table.

We begin by selecting the lower limit of the first class as a number equal to or smaller than the minimum value of the data set. Thus, in this case, since the minimum is 3.69, we select 3.50 as the lower limit of the first class. The lower limit of the second class is $3.50 + 2 = 5.50$; the lower limit of the third class is $5.50 + 2 = 7.50$, and so on. Since each data point belongs to one and only one class, the upper limit of the first class must be 5.49; the upper limit of the second class is thus 7.49, and so on. Continuing in this fashion yields the following six classes:

3.5–5.49
5.5–7.49
7.5–9.49
9.5–11.49
11.50–13.49
13.50–15.49

These classes are all mutually exclusive or nonoverlapping. Once the classes are chosen, we sort each of the 52 values into them and count the actual number of items in each class so as to obtain the class frequencies (Table 4.8a). This table represents the desired frequency distribution for our data in grouped form.

Table 4.8a Frequency Distribution of Home Mortgage Data

Serial No.	Mortgage Yields	Tally	Frequency
1	3.50–5.49	ⅲℋ /	6
2	5.50–7.49	ℋℋ ℋℋ ℋℋ ///	18
3	7.50–9.49	ℋℋ ℋℋ ℋℋ //	17
4	9.50–11.49	ℋℋ	5
5	11.50–13.49	////	4
6	13.50–15.49	//	2
Total			52

b. Relative and cumulative frequencies

The relative frequency column is obtained by dividing the frequency in each class by the total number of values n = 52; and the cumulative frequency column is obtained as shown in Table 4.8b.

Table 4.8b   Relative Frequency and Cumulative Frequency Table

Serial No.	Mortgage Yields	Frequency	Relative Frequency	Cumulative Frequency
1	3.50–5.49	6	0.12	6
2	5.50–7.49	18	0.35	$6 + 18 = 24$
3	7.50–9.49	17	0.33	$24 + 17 = 41$
4	9.50–11.49	5	0.01	$41 + 5 = 46$
5	11.50–13.49	4	0.08	$46 + 4 = 50$
6	13.50–15.49	2	0.04	$50 + 2 = 52$
Total		52	1.00	

c. Histograms

The frequency distribution can also be shown in graphical form. This is accomplished by showing continuous class intervals on the x-axis and the frequencies on the y-axis, with appropriate scales. Classes can be treated as continuous by excluding the upper limits in each class and including only the lower limits, as shown in the following.

$$5.50–7.50$$
$$7.50–9.50$$
$$9.50–11.50$$
$$11.50–13.50$$
$$13.50–15.50$$

Graphing these classes against the corresponding frequencies leads to the histogram shown in Figure 4.6.

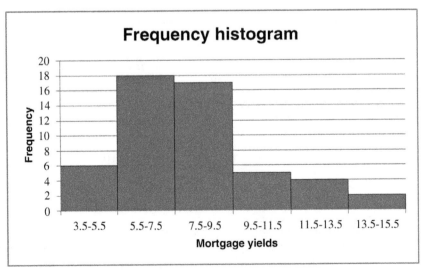

Figure 4.9   Frequency histogram.

A similar histogram, known as a relative frequency histogram (probability chart), can also be drawn using the class intervals on the x-axis and the relative frequencies (shown in Table 4.8b) on the y-axis. This is illustrated in Figure 4.10.

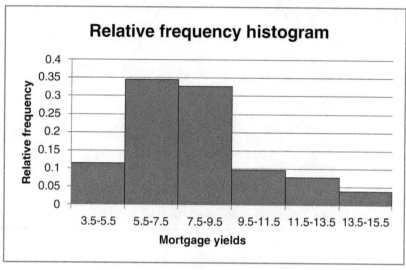

Figure 4.10   Relative frequency histogram.

d. Frequency polygon

Next comes a frequency polygon, which is obtained by plotting the midpoints of the class intervals (the average of the class limits) calculated in Table 4.8c against their frequencies and connecting the points by line segments, as shown in Figure 4.11. Clearly, this graph represents a polygon in form.

e. Ogive

The cumulative frequency curve (ogive) is obtained by plotting the upper boundary of each class against the corresponding cumulative frequency. As previously mentioned, this curve is useful in answering questions such as "How many yield values are there at or below a certain mortgage yield level"? The cumulative frequencies are shown in Table 4.8d and the cumulative frequency curve is drawn in Figure 4.12.

Table 4.8c   Frequency Table with Midpoints

Serial No.	Mortgage Yields	Midpoints	Frequency
1	3.50–5.50	4.50	6
2	5.50–7.50	6.50	18
3	7.50–9.50	8.50	17
4	9.50–11.50	10.50	5
5	11.50–13.50	12.50	4
6	13.50–15.50	14.50	2
Total			52

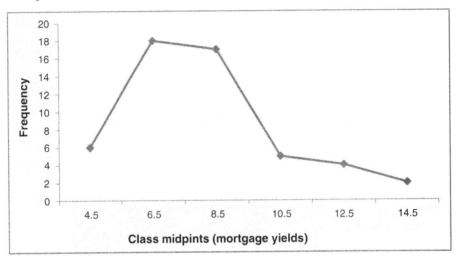

Figure 4.11   Frequency polygon.

Table 4.8d   Cumulative Frequency Values

Upper Class Boundaries	Cumulative Frequencies (Less Than or Equal to)
5.50	6
7.50	24
9.50	41
11.50	46
13.50	50
15.50	52

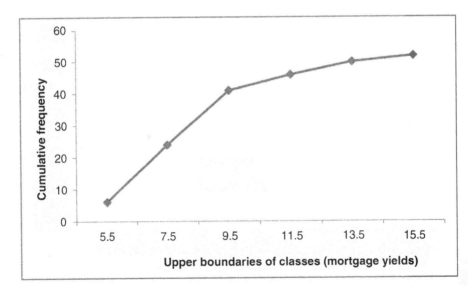

Figure 4.12   Frequency polygon

5. An undergraduate student needs a total financial commitment of $45,000 for 1 year of his college education. The student received the following financial aid package. A Pell grant of $5000 at a 3% rate, a federal loan of $18,000 at a 7% rate, family contribution of $15,000 at the rate of 11%, a grant (0% interest rate) of $4000, and the balance of $3000 is financed from individual earnings (0% interest). Find the average cost of obtaining this aid package of $45,000 for the year, ignoring opportunity cost.

SOLUTION:

The average cost of a total aid package of $45,000 for the year, ignoring opportunity cost, is

$$\overline{X} = \{5(0.03) + 18(0.07) + 15(0.11) + 4(0.0) + 3(0.0)\}/45$$
$$= 0.03\left(\frac{5}{45}\right) + 0.07\left(\frac{18}{45}\right) + 0.11\left(\frac{15}{45}\right) + 0.00\left(\frac{4}{45}\right) + 0.00\left(\frac{3}{45}\right),$$

where the weights are the proportions of the various types of loans to the total college expense of $45,000. Thus,

$$\overline{\overline{X}} = 3.06/45 = 0.068 \text{ or } 6.8\%.$$

7. Find the average growth in GDP over a period of 3 years if the GDP grew at the rate of 4% in the first year, 3% in the second year, and 5% in the third year.

SOLUTION:

As we have different rates of growth in the GDP from year to year, we have to find the average growth by the geometric mean formula. Hence,

$$\overline{X}_g = \sqrt[3]{(1 + R_1)(1 + R_2)(1 + R_3)}$$
$$= \sqrt[3]{(1 + 0.04)(1 + 0.03)(1 + 0.05)} - 1 = \sqrt[3]{(1.04)(1.03)(1.05)} - 1 = 1.039968 - 1 = 0.039968$$
or 3.9968%.

Note: If we apply the arithmetic mean formula, we obtain $[(1.04 + 1.03 + 1.05)/3] - 1 = 1.04 - 1 = 0.04$ or 4%, which is a wrong answer because, if we apply this annual rate of growth, the actual amount at the end of the 3 years will be inflated.

9. It is reported that the total medical care cost in the United States over the period of 10 years from 1996 to 2005 grew at a rate of 42%. Find the annual average growth rate (medical care cost inflation) during this 10-year period.
(*Hint:* Consider $(1+R)^{10} = 1.42$ and find R.)

SOLUTION:

Let the medical care cost in the year 1996 be X_1 and let R be the annual average growth rate. Then, the medical care cost in the year 2005, X_{10}, that is, after 10 years, will be $X_1 (1+R)^{10} = X_{10}$. Then, the rate of growth over the period of 10 years is $X_{10}/X_1 = (1 + R)^{10} = 1.42$, which is given in the problem. Solving for R, we get $1 + R = 1.42^{\frac{1}{10}} = 1.42^{0.1} = 1.035688$ or $R = 1.035688 - 1 = 0.035688$ or 3.5688%, approximately.

11. Suppose that the mean GPA of a sample of 280 MBA students is 3.57 with a standard deviation of 0.23. Find the proportion of students with GPA's ranging from 3.30 to 3.84.

SOLUTION:

The Z- score corresponding to the GPA value of 3.30 is $(3.30 - 3.57)/0.23 = -1.17$, and the Z-score corresponding to a GPA of 3.84 is $(3.84 - 3.57)/0.23 = 1.17$.

Thus, we see that the two GPA values of 3.30 and 3.84 are within 1.17 standard deviations of the mean. Thus, according to Chebysheff's theorem, at least $(1 - (1/K^2)) = 1 - (1/1.17^2) = 1 - (1/1.3689) = 1 - 0.7305 = 0.2695$ or 26.95% of students are within a GPA of 3.30 and 3.84.

13. The Bureau of Labor Statistics reported that the mean price per gallon of unleaded regular gas for the year 2011 was $3.53 per gallon with a standard deviation of $0.25 per gallon. Assuming that the distribution of gas price is mound-shaped, find a gas price per gallon range centered around the mean within which 95% of the prices would fall.

SOLUTION:

$\overline{X} - 2s = 3.53 - 2(0.25) = 3.53 - 0.50 = 3.03$

$\overline{X} + 2s = 3.53 + 2(0.25) = 3.53 + 0.50 = 4.03.$

Therefore, by the empirical rule, 95% of the gas prices fell between $3.03 and $4.03 per gallon in the year 2011.

15. The following is a sample of P/E ratios of 10 international companies.

15, 7, 26, 16, 11, 9, 5, 14, 11, 28

Find the 85th percentile and 3rd quartile for this data set.

SOLUTION:

Using the general location formula for any percentile, $P_i = i(n + 1)/100$ for $i = 1, 2, 3, \ldots, 9$, the 85th percentile is located at the $P_{85} = (85/100)(10 + 1) = 9.35$th term, after arranging the data in ascending order of magnitude. The arranged data are as follows:

$$5, 7, 9, 11, 11, 14, 15, 16, 26, 28$$

Thus, the 85th percentile is found by interpolation as follows: 9th item value + 2(the difference between the 9th and 10th item values)

$$= 26 + 0.35(28 - 26) = 26 + 0.35(2) = 26.7.$$

The 75th percentile is located at $P_{75} = (75/100)(10 + 1) = 8.25$th term. Then, by interpolation, the 75th percentile is the 8th item value + 0.25 $(26 - 16) = 16 + 2.5 = 18.50.$

17. The following are the temperature forecasts, high/low, for 10 days, June 5–June 15, in 2012 in the greater Hartford area of Connecticut, as reported on weather.com.

63/52	72/53	75/56	76/59	80/64	85/65	85/67	88/67	83/63	80/59.

Find the modes of high temperature as well as low temperature forecasts.

SOLUTION:

High temperature modes: 85 (occurs twice), 80 (occurs twice)

Low temperature modes: 59 (occurs twice) 67 (occurs twice)

EXCEL APPLICATIONS

19. The accompanying sample data set shows the external debt of countries in the year 2014, in millions of dollars. *Data file:* External debt.xlsx

Find the median debt of the above countries and interpret its value.

SOLUTION:

Enter the data in an Excel spreadsheet, and highlight the data. Click on the Icon

"Z below A" on the top right-hand side of the ribbon and select the option, "smallest to largest." This arranged data in the column can be transposed to a row by highlighting the data, right click, select "copy" and choose a cell where you want to start the row. Then "right click" and click on "paste" and choose the option "transpose".

You see the following debt figures in ascending order of magnitude:

Country (2014) and external debt						
Algeria	Nigeria	Bangladesh	Egypt	Pakistan	Vietnam	Venezuela
4,872	22,010	33,200	55,860	62,330	68,050	69,660
Malaysia	Indonesia	Turkey	India	Mexico	Russia	China
109,300	278,500	407,100	425,300	438,400	683,600	894,900

Since there is an even number of countries (14), the median is calculated as the mean of the middle two values of the external debt series. Thus, the mean of $n/2 = 14/2 = 7$th value and $(n/2) + 1 = (14/2) + 1 = 7 + 1 = 8$th value must be calculated. That is, the median is $(69,660 + 109,300)/2 = 178,960/2 = \$89,480$ millions.

There are seven countries (Algeria, Nigeria, Bangladesh, Egypt, Pakistan, Vietnam, and Venezuela) having an external debt equal to or less than $89,480 millions; and another seven countries (Malaysia, Indonesia, Turkey, India, Mexico, Russia, and China) having an external debt above $89,480 millions.

(*Note:* The data can be arranged in ascending order in a column as follows: Open the *data file* External debt.xlsx that contains the data on external debt and sort the values in ascending order using Excel. To accomplish this, first highlight the data on external debt; press *control*, *shift* and *down arrow* simultaneously. Click on the icon "Z below A" on the top right and then click on the option, "smallest to largest".)

21. Find a measure of the degree of skewness for the following set of employment data. Comment on the nature of its skewness. *Data file:* Employment2012.xlsx

SOLUTION:

Go to Excel data sets and open the *Employment2012.xlsx* file. Click on *Data* in the menu bar, click on *Data Analysis,* and select "Descriptive Statistics." Follow the dialog box, check the "Summary Statistics" box, and click OK. Then you will find the summary statistics, which includes

Mean = 287.86

Median = 197

Mode = 149

Standard deviation = 295.85.

Based on this information, a relative measure (coefficient) of skewness is (mean − mode)/standard deviation = (287.86 − 149)/295.85 = 0.469.

Comment: Since the measure of skewness is positive, we can say that the distribution is "positively skewed" or we can say that "the distribution has an elongated right tail."

Note: Excel software also gives a measure of skewness, but it is based on a different formula.

23. Using Excel, find the 40th percentile, median, and 3rd quartile of the following data on average SAT scores of students in undergraduate B-Schools as reported by the May 8, 2006, issue of *Business Week* magazine. *Data file:* SAT scores.xlsx

Interpret your values obtained.

SOLUTION:

Note that there are 47 SAT scores and they are sorted using Excel in ascending order in the following:

Enter the data in an Excel spreadsheet, highlight the data. Click on the Icon

"Z below A" on the top right-hand side of the ribbon and select the option, "smallest to largest". These arranged data in the column can be transposed to a row by highlighting the data, right click, select "copy," and choose a cell where you want to start the row. Then "right click" and click on "paste" and choose the option "transpose."

SI#	1	2	3	4	5	6	7	8	9	10	11	12	13	14	15
	1150	1159	1162	1167	1169	1176	1185	1189	1190	1191	1197	1199	1212	1218	1220
SI#	16	17	18	19	20	21	22	23	24	25	26	27	28	29	30
	1225	1230	1231	1246	1260	1260	1263	1273	1283	1283	1291	1300	1300	1302	1304
SI#	31	32	33	34	35	36	37	38	39	40	41	42	43	44	45
	1314	1315	1319	1330	1332	1333	1339	1341	1350	1351	1360	1379	1381	1422	1448
SI#	46	47													
	1451	1477													

The location of the 40th percentile is given by the formula, $P_{40} = [40(n + 1)]/100$ or

$$P_{40} = \frac{40}{100}(47 + 1) = \frac{2}{5}(48) = \frac{96}{5} = 19.2.$$

Thus, the location of the 40th percentile is between the 19th and 20th values in the sorted data.

Then, it is interpolated as $1246 + 0.2(1260 - 1246) = 1246 + 0.2(14) = 1246 + 2.8 = 1248.8$.

Interpretation: 40% of the scores are at or below a score of approximately 1248.8.

Median is the 50th percentile and its location is given by $P_{50} = \frac{50}{100}(47 + 1) = \frac{1}{2}(48) = 24$.

Therefore, the 24th value in the sorted data is the 50th percentile or median, which is 1283.

Interpretation: 50% of the scores are at or below a score of 1283.

Third quartile is the same as the 75th percentile and its location is given by

$$P_{75} = \frac{75}{100}(47 + 1) = \frac{3}{4}(48) = 36.$$

Therefore, the third quartile is the 36th value in the sorted data, which is 1333.

Interpretation: 75% of the scores are at or below 1333.

25. Compare the relative variability of exports and imports of the United States from 1986 to 2011, as given in the following table. *Data file:* Exports and Imports.xlsx

Year	Exports	Imports
1986	320.3	452.9
1987	363.8	508.7
1988	443.9	554
⋮	⋮	⋮
2009	1,583.00	1,974.60
2010	1,839.00	2,356.70
2011	2,087.60	2,665.80

SOLUTION:

Let X stand for Exports and Y stand for Imports.

Go to Excel data sets and open the *Exports and Imports.xlsx* file. Click on *Data* in the menu bar, click on *Data Analysis* and select "Descriptive Statistics." Follow the dialog box, check the "Summary Statistics" box, and click OK. Then you will find the summary statistics as shown in the following table:

Mean	1003.734615
Standard error	94.71494171
Median	954.15
Mode	#N/A
Standard deviation	482.953336
Sample variance	233,243.9248
Kurtosis	−0.237981489
Skewness	0.72121022
Range	1767.3
Minimum	320.3
Maximum	2087.6
Sum	26,097.1
Count	26

Coefficient of variation of exports (X) is $(482.95/1003.73) \times 100\% = 48.12\%$

Similarly, following the above procedure of using *Data Analysis to* Imports data (Y), we obtain the summary statistics as shown in the following table:

Y	
Mean	1334.430769
Standard error	139.0631136
Median	1183.55
Mode	#N/A
Standard deviation	709.0855298
Sample variance	502,802.2886
Kurtosis	−1.086694587
Skewness	0.504964486
Range	2212.9
Minimum	452.9
Maximum	2665.8
Sum	34,695.2
Count	26

Coefficient of variation is $(709.09/1334.43) \times 100\% = 53.14\%$

Comment: Imports show slightly higher variability $(CV = 53.14\%)$ than exports $(CV = 48.12\%)$ during the period 1986–2011.

27. Use Excel (Data Analysis tool) to answer the following questions for the data on unemployment rate for both males and females. *Data file:* UnemploymentMF2013.xlsx

   a. Find the mean, median, and the mode for each data series.

   b. Based on the values of the mean, median, and mode, can you comment on the shape of each distribution? More specifically, does it look like bell-shaped, skewed to the right, or skewed to the left?

   c. Find the variance, standard deviation, and the coefficient of variation for each data series.

SOLUTION:

a.

> **Step 1:** Open *data file* "UnemploymentMF2013.xlsx" in Excel data sets.
> **Step 2:** Click on "Data" on the menu bar and select "Data Analysis" on the right extreme side of the ribbon.
>
> If you don't see the "Data Analysis," go to "File" and select the "Add-Ins" option and click on "Analysis Tool Pak" and press "OK".
> **Step 3:** Select the "Descriptive Statistics" from the Data Analysis menu.

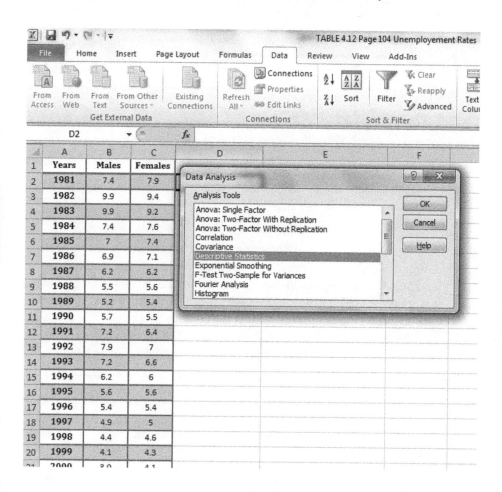

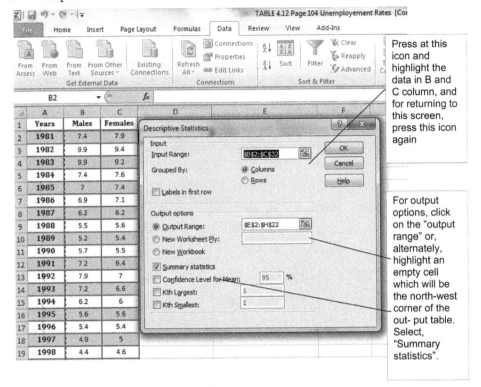

An easy way of highlighting the input range is to highlight the first couple of cell values (input) in B column and keep the buttons, "Shift" and "Ctrl" depressed at the same time with your left-hand fingers while you press the "down arrow" key with you right-hand finger. This action will highlight the entire data only in the column B.

Press "OK" on the screen above. You will have the descriptive statistics output table in the Excel sheet as shown in the following table:

Unemployment Rate Males		Unemployment Rate Females	
Mean	6.490322581	Mean	6.209677
Standard error	0.336054106	Standard error	0.268783
Median	6.1	Median	5.6
Mode	7.4	Mode	5.4
Standard deviation	1.871070075	Standard deviation	1.496519
Sample variance	3.500903226	Sample variance	2.23957
Kurtosis	−0.128460489	Kurtosis	−0.58231
Skewness	0.859740404	Skewness	0.664836
Range	6.6	Range	5.3
Minimum	3.9	Minimum	4.1
Maximum	10.5	Maximum	9.4
Sum	201.2	Sum	192.5
Count	31	Count	31

b. Both of them are slightly positively skewed, meaning that they have an elongated right tail.

c. Variance of the unemployment rate for males is 3.5 and for females it is 2.24.

Standard deviation of the unemployment rate for males is 1.87 and for females it is 1.50.

Coefficient of variation of the unemployment rate for males, computed as the standard deviation divided by the mean, is $1.87/6.49 = 0.29$ or 29% and for females it is computed as $1.50/6.2 = 0.24$ or 24%.

Chapter 5

Probability and Applications

5.2 SOME USEFUL DEFINITIONS

Random experiment: a process that has an unknown outcome or outcomes that are known only after the process is completed.

   Event: an outcome of a random experiment.

   Sample space: a collection of *all possible* outcomes of a random experiment (denoted S).

   Simple event: a single point within the sample space.

   Compound event: a collection of simple events.

5.3 PROBABILITY SOURCES

5.3.1 Objective Probability

Classical (objective) probability: if a random experiment has N equally likely outcomes and N_A of them are favorable to event A, then the probability of event A is

$$P(A) = N_A/N.$$

Relative frequency as a probability: if a random experiment is repeated a large number (n) of times and event A is observed in N_A of these n repetitions, then the *relative frequency* of event A is n_A/n. Moreover, if n_A/n approaches some long-run stable value P(A) as $n \to \infty$, then $\lim_{n\to\infty} n_A/n = P(A)$ is called the *probability of event A*.

5.4 SOME USEFUL DEFINITIONS INVOLVING SETS OF EVENTS IN THE SAMPLE SPACE

Suppose A and B are events within the sample space S.

Solutions Manual to Accompany Introduction to Quantitative Methods in Business: With Applications Using Microsoft® Office Excel®, First Edition. Bharat Kolluri, Michael J. Panik, and Rao N. Singamsetti.
© 2017 John Wiley & Sons, Inc. Published 2017 by John Wiley & Sons, Inc.

Union of events A and B: the set of simple events that are either in A or in B or in both A and B.

Intersection of events A and B: the set of simple events that are common to events A and B.

Complement of event A (denoted $\overline{A}$): consists of all simple events within S that are *not* in A.

Mutually exclusive events: events A and B are mutually exclusive if $A \cap B = \phi$. Thus, the occurrence of A precludes the occurrence of B and vice versa.

5.5   PROBABILITY LAWS

For events A and B within S:

General additional rule: $P(A \cup B) = P(A) + P(B) - P(A \cap B)$. Here, $P(A)$ and $P(B)$ are *marginal probabilities* and $P(A \cap B)$ is a *joint probability*.

Special additional rule: if $A \cap B = \phi$, then $P(A \cup B) = P(A) + P(B)$.

5.5.2   Rule of Complements

According to the *Rule of Complements*, $P(\overline{A}) = 1 - P(A)$, that is, the probability that event A *does not* occur equals 1 minus the probability that event A *does occur*. This is because $A \cup \overline{A} = S$ and $P(S) = 1$.

5.5.3   Conditional Probability

Suppose an event B has definitely occurred. What is the probability that an event A has also occurred? The *probability of A given B* is called the *conditional probability* of A given B and written $P(A|B)$, where " | " is read "given." In general,

$$P(A|B) = \frac{P(A \cap B)}{P(B)}, \quad P(B) \neq 0,$$

$$P(B|A) = \frac{P(A \cap B)}{P(A)}, \quad P(A) \neq 0.$$

5.5.4   General Multiplication Rule (Product Rule)

From the definition of the conditional probabilities $P(A|B)$ and $P(B|A)$, we can write the *general multiplication rule* as

$$P(A \cap B) = P(A|B) \cdot P(B)$$
$$= P(B|A) \cdot P(A)$$

since $P(A \cap B)$ appears in the numerator of both of these conditional probabilities.

5.5.5 Independent Events

Two events A and B are *independent* if the occurrence of one of them in no way affects the probability of occurrence of the other. Hence, A and B are independent if and only if $P(A|B) = P(A)$ and $P(B|A) = P(B)$. For independent events, we have the *special multiplication rule*:

$$P(A \cap B) = P(A) \cdot P(B).$$

What is the distinction between events that are mutually exclusive and events that are independent? Mutually exclusive events *cannot occur together*, that is, $A \cap B$ cannot occur. Thus, $A \cap B = \phi$ and thus $P(A \cap B) = 0$. Independent events *can occur together*: It is just that when one occurs, it does not affect the probability of occurrence of the other. Hence, $A \cap B \neq \phi$ and thus $P(A \cap B) \neq 0$. In fact, $P(A \cap B) = P(A) \cdot P(B)$.

5.5.6 Probability Tree Approach

This approach is useful when we have multiple trials of a random experiment. For instance, suppose we have two possible outcomes for our random experiment (events A and B) and that $P(A) + P(B) = 1$. Suppose also that three separate trials are conducted. From an initial node, the tree starts with two branches representing the occurrence of either A or B on the first trial (Figure 5.1). From each of these two nodes, we branch again for trial two. And on trial three, we branch from each of the four nodes determined on trial two. We finish with eight terminal nodes. For instance, under three independent trials, we branch as indicated in Figure 5.1.

5.6 CONTINGENCY TABLE

A *contingency table* is an array of data arranged in rows and columns. It is a useful device for determining probabilities. For example, given two row categories (R_1 and R_2) and column categories (C_1 and C_2), Table 5.1 houses the distribution of $n = 130$ data values. Hence, we may compute

$$P(R_1 \cap C_1) = \frac{30}{100}, \quad P(C_2) = \frac{90}{130},$$

$$P(C_1|R_2) = \frac{P(C_1 \cap R_2)}{P(R_2)} = \frac{10/130}{60/130} = 10/60, \text{ and so on}$$

Table 5.1 can be transformed into a *probability table* (Table 5.2) by dividing each of its entries by 130.

Table 5.1   Contingency Table

	C_1	C_2	Total
R_1	30	40	70
R_2	10	50	60
Total	40	90	130

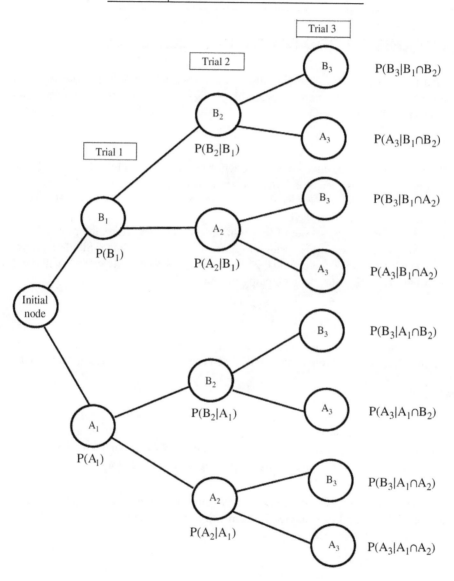

$$P(A_1 \cap B_2 \cap B_3) = P(A_1)P(B_2)P(B_3),$$
$$P(B_1 \cap A_2 \cap B_3) = P(B_1)P(A_2)P(B_3), \text{ and so on}$$

Figure 5.1   Probability tree.

Table 5.2   Probability Table

	C_1	C_2	Total
R_1	30/130	40/130	70/130
R_2	10/130	50/130	60/130
Total	40/130	90/130	1

SOLUTIONS TO ODD-NUMBERED EXERCISES

1. Find the probability of observing three heads when you toss a coin three times or three coins once. Also, find the probability of getting four heads if you toss a coin four times or four coins once.

SOLUTION:

1. When you toss a coin three times, there will be $2^3 = 8$ possibilities. Here, 2 refers to two possibilities, H (head) or T (tail), when a coin is tossed once and 3 refers to the number of times the coin is tossed. Here the possibilities are as follows: HHH, HHT, HTH, HTT, THT, TTH, THH, and TTT. Thus, we can see that there is only one possibility for three heads out of the eight possibilities. Therefore, the probability of observing three heads is 1/8, using the classical rule of probability. Similarly, if you toss a coin four times, there will be $2^4 = 16$ possibilities and only one of them corresponds to HHHH. Thus, the probability of four heads is

$$\frac{1}{2^4} = \frac{1}{16}$$

3. Find the probability of picking any card at random with a numerical value less than 10 in a deck of 52 playing cards. Consider an ace as having a numerical value of 1.

SOLUTION:

3. Count an ace as one. Note that there are 1–10 numerical values for cards in each of four suites of playing cards. Thus, there will be $4 \times 9 = 36$ cards among 52 cards, which satisfy the description of the favorable event of a card with a numerical value less than 10. Thus, the probability of picking a card at random with a numerical value less than 10 is $36/52$.

5. Find the probability of selecting three even-numbered balls without replacement out of a collection of 59 balls numbered from 1 to 59.

SOLUTION:

5. There are 29 even numbers from 1 to 59: 2,4,6,8,10,12,14,16,18,20,22,24,26,28,30, 32,34,36,38,40,42,44,46,48,50,52,54,56, and 58.
$29/59 \times 28/58 \times 27/57 = 21,924/195,054 = 0.1124$ (rounding-off to four decimal places) or 11.24%.

7. Suppose 60% of MBA graduates read the *Wall Street Journal*, 50% read the *New York Times*, and 30% read both. Find the probability that a randomly selected MBA

graduate reads at least one of them. Also, find the probability that this graduate reads neither of them.

SOLUTION:

7. Let A stand for the event "MBA graduates read the *Wall Street Journal*," and let B represent the event "MBA graduates read the *New York Times*." Then,

$$P(A \cup B) = P(A) + P(B) - P(A \cap B) = 60\% + 50\% - 30\% = 110\% - 30\% = 80\%$$

Therefore, 80% of the MBA graduates read either the *Wall Street Journal* or the *New York Times*, or both. The probability that this graduate reads neither of them is

$$1 - P(A \cup B) = 1 - 80\% = 1 - 0.80$$
$$= 0.2 \text{ or } 20\% \text{ using the probability rule of complements.}$$

9. In 2014, the U.S. Senate was made up of 45 Republicans, 53 Democrats, and 2 Independents. Find the probability of selecting two Republican senators sequentially (one after another) with and without replacement.

SOLUTION:

9. The probability of selecting two Republican senators with replacement using the multiplication rule of probability is

$$45/100 \times 45/100 = 0.2025 \text{ or } 20.25\%.$$

The probability of selecting two Republican senators without replacement is

$$45/100 \times 44/99 = 0.2 \text{ or } 20\%.$$

11. Suppose a college basketball team of five players contains three graduating seniors, find the probability of selecting three graduating seniors sequentially, assuming replacement.

SOLUTION:

11. In a basketball team of five players, there are three graduating seniors. The probability of selecting three graduating seniors sequentially (with replacement) is

$$1/5 \times 1/5 \times 1/5 = 0.008 \text{ or } 0.8\%.$$

13. The following contingency table relates data on educational attainment by people of age 25 or older to ethnicity in the United States in the year 2000 (*Source:* U.S. Census Bureau, Statistical Abstract of the United States: 2002).

Table 5.3   Educational Attainment versus Ethnicity (2000)

	High School Graduate or Less (C_1)	Some College, No Degree (C_2)	Under- graduate Degree (C_3)	Master's Degree and Higher (C_4)	Total
White (R_1)	71,327	37,355	25,443	12,942	147,067
Black (R_2)	11,360	5,370	2,284	1,022	20,036
Others (R_3)	3,275	1,731	2,048	1,073	8,127
Total	85,963	44,456	29,775	15,036	175,230

Answer the following questions based on the data provided in the table above.

a. What is the probability of selecting a black person?

b. What is the probability of selecting a white person or a person with an educational level lower than an undergraduate degree?

c. What is the probability that a randomly selected black person has a master's degree and higher?

d. Two people are selected at random and without replacement. What is the probability that both are "high school graduates, or less?" What is your answer if sampling is done with replacement?

e. Are ethnicity and educational attainment statistically independent?

SOLUTION:

13. *Notation:* Let R_i denote row i, where $i = 1$ for white, $i = 2$ for black, and $i = 3$ for others. Likewise, let C_j represent column j, where $j = 1$ for high school graduate or less, $j = 2$ for some college, no degree, $j = 3$ for undergraduate degree, and $j = 4$ for master's degree and higher.

a. $P(R_2) = \dfrac{\text{Black persons}}{\text{Total number}} = \dfrac{20,036}{175,230} = 0.1143.$

b. $P(R_1 \cup (C_1 \cup C_2)) = P(R_1) + P(C_1 \cup C_2) - P(R_1 \cap (C_1 \cup C_2))$

$$= \frac{147,067}{175,230} + \left(\frac{85,963 + 44,456}{175,230} \right) - \left(\frac{71,327 + 37,355}{175,230} \right)$$

$$= 0.8393 + 0.7443 - 0.6202$$

$$= 0.9634.$$

c. $P(C_4 | R_2) = \dfrac{P(C_4 \cap R_2)}{P(R_2)} = \dfrac{1022/175,230}{20,036/175,230}$

$\qquad = \dfrac{0.0058}{0.1143} = 0.0507.$

d. Let S_1 and S_2 represent two people in the high school graduate or less category. Then,

$$P(S_1 \cap S_2) = P(S_1).P(S_2|S_1)$$

$$= \frac{85,963}{175,230} \times \frac{85,962}{175,229}$$

$$= 0.24066 \text{ (without replacement)}$$

If sampling is done with replacement,

$$P(S_1 \cap S_2) = P(S_1).P(S_2) = \frac{85,963}{175,230} \times \frac{85,963}{175,230} = 0.2406611.$$

e. To find whether ethnicity and educational attainment are statistically independent, test if, for instance,

$$P(C_1|R_1) = P(C_1). \text{ Here } P(C_1|R_1) = \frac{71,327}{147,067} = 0.4849 \text{ and } P(C_1) = \frac{85,963}{175,230}$$

$$= 0.49057.$$

Since these two probabilities are not equal ($0.4849 \neq 49057$), we conclude that ethnicity and educational attainment are not independent.

15. The following contingency table gives locations of different Wal-Mart stores in the United States in the year 2002 (*Source:* Wal-Mart 2002 Annual Report).

Table 5.4

	Discount Stores (C_1)	Supercenters (C_2)	SAM's Club (C_3)	Neighborhood Markets (C_4)	Total
Northeast (R_1)	243	83	60	0	386
Midwest (R_2)	460	266	146	0	872
South (R_3)	591	790	238	48	1667
West (R_4)	274	119	81	1	475
Total	1568	1258	525	49	3400

Based on these data, answer the following questions:

a. What is the probability of selecting a store in the Northeast?

b. What is the probability of selecting a store in the Midwest or a discount store nationwide?

c. If a store in the South is selected at random, what is the probability that this store is Sam's Club?

d. Two stores are selected at random and without replacement. What is the probability that both are in the West?

SOLUTION:

15. Let C_j denote column j, j = 1,2,3,4, and R_i stand for row i, i = 1,2,3,4.

a. The probability of selecting a store in the northeast is $P(R_1) = 386/3400 = 0.113$ (11.35%).

b. $P(R_2 \cup C_1) = P(R_2) + P(C_1) - P(R_2 \cap C_1)$
$$= 872/3400 + 1568/3400 - 460/3400$$
$$= 0.2565 + 0.4612 - 0.1353$$
$$= 0.5824 \, (58.24\%).$$

c. $P(C_3|R_3) = \frac{P(C_3 R_3)}{P(R_3)} = (238/3400)/(1667/3400) = 238/1667 = 0.1428 (14.28\%).$

d. Let S_1 and S_2 refer to two stores in the West selected at random.
$$P(S_1 \cap S_2) = P(S_1) \cdot P(S_2|S_1)$$
$$= (475/3400)(474/3399) = 0.1397 \times 0.1395 = 0.0195 (1.95\%).$$

EXCEL APPLICATIONS

17. Find solutions to Exercise 15 using Excel.

SOLUTION:

17. Reproduce the contingency table given in Exercise 15 on an Excel spreadsheet. Then divide all the frequencies (numbers) by the grand total of 3400 and form a similar table, which may be called a *probability table*. The interior probabilities are joint probabilities and the row end/column end probabilities are called marginal probabilities. Then the answers to the questions can be found as described in the Excel table below.

	A	B	C	D	E	F	G	H
1		Discount Stores (C1)	Supercenters (C2)	SAM's Club (C3)	Neighborhood(C4)	Total		
2	Northeast (R1)	243	83	60	0	386		
3	Midwest (R2)	460	266	146	0	872		
4	South (R3)	591	790	238	48	1667		
5	West (R4)	274	119	81	1	475		
6	Total	1568	1258	525	49	3400		
7								
8	Probability table							
9								
10	Northeast (R1)	0.071470588	0.024411765	0.017647059	0	0.113529		
11	Midwest (R2)	0.135294118	0.078235294	0.042941176	0	0.256471		
12	South (R3)	0.173823529	0.232352941	0.07	0.014117647	0.490294		
13	West (R4)	0.080588235	0.035	0.023823529	0.000294118	0.139706		
14	Total	0.461176471	0.37	0.154411765	0.014411765	1		
15								
16	Ans a.	P(R1) = 0.113529	⟶		CellF10 Marginal probability			
17								
18	Ans b.	P(R2) + P(C1) - P(R2∩C1) = 0.256471 + 0.461176471 - 0.135294118 = 0.582353						
19								
20	Ans c.	P(C3|R3) = P(C3∩R3)/P(R3) = 0.07/0.490294 = 0.142771						
21								
22	Ans d.	P(selecting one store in R4) * P(second store in R4|the first is not replaced) = CellF13*(474/3399).						
23		0.139706*0.139453 = 0.0194824. Note that the second probability is from the contingency table.						
24								
25								

Chapter 6

Random Variables and Probability Distributions

6.2 PROBABILITY DISTRIBUTION OF A DISCRETE RANDOM VARIABLE X

A random variable X is *discrete* if it takes on a finite or countably infinite number of values. The display of all possible values of a discrete random variable X along with their associated probabilities is called a *discrete probability distribution*. It has two properties:

$$0 \leq P(X_i) \leq 1 \quad \text{and} \quad \sum P(X_i) = 1.$$

(Table 6.1). The graphical depiction of a discrete probability distribution is called a *probability mass function* (Figure 6.1).

Table 6.1   Discrete Probability Distribution

X	P(X)
X_1	$P(X_1)$
X_2	$P(X_2)$
$\vdots$	$\vdots$
X_n	$P(X_n)$
Total	1.0

Solutions Manual to Accompany Introduction to Quantitative Methods in Business: With Applications Using Microsoft® Office Excel®, First Edition. Bharat Kolluri, Michael J. Panik, and Rao N. Singamsetti.
© 2017 John Wiley & Sons, Inc. Published 2017 by John Wiley & Sons, Inc.

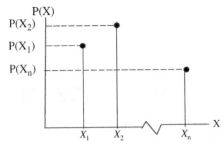

Figure 6.1 Probability mass function.

6.3 EXPECTED VALUE, VARIANCE, AND STANDARD DEVIATION OF A DISCRETE RANDOM VARIABLE

For a discrete random variable X, the *expected value* of X is denoted as

$$\mu = E(X) = \sum X_i P(X_i).$$

Here, $E(X)$ serves as the *mean* of a discrete probability distribution.

The *variance* of a discrete random variable X is expressed as

$$\sigma^2 = \sum (X_i - \mu)^2 P(X_i) \quad \text{(long formula)}$$

or as

$$\sigma^2 = \sum X_i^2 P(X_i) - \mu^2 \quad \text{(short formula)}.$$

The *standard deviation* of X is then simply

$$\sigma = \sqrt{\sigma^2} = \sqrt{\sum (X_i - \mu)^2 P(X_i)} \quad \text{(long formula)}$$
$$= \sqrt{\sum X_i^2 P(X_i) - \mu^2}. \quad \text{(short formula)}$$

Here, σ serves as a measure of the *average variability* of the X_i values about the mean μ.

6.3.1 Some Basic Rules of Expectation

For X a discrete random variable with A and B constants:

1. $E(A) = A$.
2. $E(A + BX) = A + BE(X)$.
3. For discrete random variables $X_1, \ldots, X_K$ whose individual expectation $E(X_K)$ exists,

$$E\left(\sum_{j=1}^{K} X_j \right) = \sum_{j=1}^{K} E(X_j).$$

A useful decision-making tool is the *payoff table* consisting of, say, *net profit values* corresponding to various combinations of situations/actions. Using probabilities corresponding to each situation, we can determine the *expected payoff values* for each action. Hence, the *optimal decision is to select the action that gives the maximum expected payoff.*

Alternatively, we can define the *opportunity loss* as the difference between the maximum profit that would be realized if situation j occurs and the net profit level appearing in the payoff table. (Note that opportunity loss must always be nonnegative.) Once the various opportunity losses are determined, we find the expected opportunity loss corresponding to each possible action. Then, the *optimal decision involves selecting the action that minimizes expected loss.* The expected payoff versus expected loss approaches must result in the same action taken. (The reader is encouraged to review Examples 6.9 and 6.10 in the main text.)

6.3.2   Some Useful Properties of the Variance of X

For X a discrete random variable with A and B constants, the variance of X can generally be written as

$$\sigma^2 = V(X) = E[X - E(X)]^2$$
$$= \sum (X_i - E(X))^2 P(X_i) \quad \text{(long formula)}$$

or

$$\sigma^2 = V(X) = E(X^2) - E(X)^2$$
$$= \sum X_i^2 P(X_i) - E(X)^2 \quad \text{(short formula)}$$

Then,

1. $V(A) = 0$.
2. $V(A + BX) = 0 + B^2 V(X) = B^2 \sigma^2$.
3. For $Z = \frac{X - \mu}{\sigma}$,
   a. $E(Z) = 0$.
   b. $V(Z) = 1$.

6.4   CONTINUOUS RANDOM VARIABLES AND THEIR PROBABILITY DISTRIBUTIONS

A probability distribution is *continuous* if its associated random variable X is continuous, that is, X can assume any value over an interval form a to b. The interval can be open or closed.

The probability distribution of a continuous random variable can be described by its *probability density function* f(x). Here, f(x) is nonnegative, the area under the curve of f(x) over an interval is interpreted as *probability*, and the total area under f(x) over its range of definition must be unity. Now, an *event* is an interval on the X-axis and

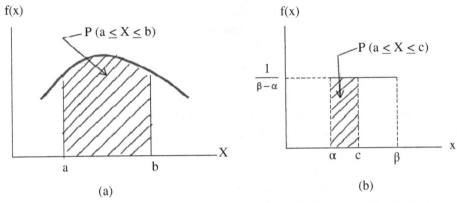

Figure 6.2 (a) Probability as an area under f (x) over [a,b]. (b) Uniform probability distribution.

$$P(a \leq X \leq b) = \int_a^b f(x)d(x)$$

(see Figure 6.2a). (*Note:* P(a ≤ X ≤ b) = P(a < X < b) since P(a) = P(b) = 0.) A particular type of continuous probability distribution is the *uniform probability distribution* (Figure 6.2b).

6.5 A SPECIFIC DISCRETE PROBABILITY DISTRIBUTION: THE BINOMIAL CASE

6.5.1 Binomial Probability Distribution

Let a random experiment be characterized as a *simple alternative* (or *binomial random experiment*), that is, there are only two mutually exclusive outcomes—*success* or *failure*.

The features of a binomial experiment are as follows:

i. Two possible outcomes (success or failure).

ii. The discrete random variable X is defined as the number of successes observed in n independent and identical trials of the simple alternative experiment.

iii. The probability of a success p is constant from trial to trial (think of sampling with replacement).

It is important to note that the order in which the successes are obtained is irrelevant. In fact, the total number of ways that, say, r successes can be obtained in the n independent trials of a binomial experiment is given by the formula

$$_nC_r = \frac{n!}{r!(n-r)!}, \quad r \leq n.$$

Table 6.2   Binomial Probability Distribution

X	P(X; n p)
0	P(0; n,p)
1	P(1; n,p)
2	P(2; n,p)
⋮	⋮
n	P(n; n, p)
Total	1.0

Here $_nC_r$ is read as "the number of combinations of n distinct items taken r (the number of successes) at a time.

So if n stands for the number of independent trials of a simple alternative experiment, p denotes the probability of a success, $q = 1 - p$ represents the probability of a failure, then the probability of getting r successes in the n trials is

$$P(r; n, p) = {_nC_r}\, p^r q^{n-r},$$

the *binomial probability function.*

To obtain the binomial probability distribution, let X depict the number of successes in the n independent trials. Then as X varies from 0 to n, we generate the binomial probability distribution (Table 6.2), where the *general binomial probability function* appears as

$$P(X; n, p) = {_nC_X} p^X q^{n-X}$$
$$= \frac{n!}{X!(n-X)!} p^X (1-p)^{n-X}, \quad X = 0, 1, 2, \ldots, n.$$

One need not always use the binomial probability function to calculate binomial probabilities. Table 6.A.1 of the main text contains binomial probabilities for selected values of the *binomial parameters* n, p.

6.5.2 Mean and Standard Deviation of the Binomial Random Variable

The mean and standard deviation of a binomial random variable X can readily be calculated as

$$E(X) = np = \mu$$

and

$$V(X) = np(1-p) = \sigma^2,$$

respectively. The standard deviation is then

$$\sqrt{V(X)} = \sqrt{np(1-p)} = \sigma.$$

6.5.3 Cumulative Binomial Probability Distribution

A *cumulative binomial probability*, such as the probability of obtaining "at most" a given number of successes in a specific number of trials in a binomial experiment, can be calculated as

$$P(X \le x) = \sum_{0}^{x} P(X; n, p),$$

where $P(X; n, p)$ is the binomial probability function for X. To find $P(X > x)$, we simply calculate $P(X > x) = 1 - P(X \le x)$. Table 6.A.2 of the main text presents, for specific values of n and p, cumulative binomial probabilities.

SOLUTIONS TO ODD-NUMBERED EXERCISES

1. Determine which of the following are legitimate discrete probability distributions. Justify your answers.

a.	X	P(X)	b.	X	P(X)	c.	X	P(X)
	−3	0.8		1	0.4		0	0.2
	2	0.2		3	0.3		2	0.3
	4	0.1		5	0.3		3	0.3
	10	−0.1		7	0.2		4	0.2

SOLUTION:

a. It is not a probability distribution since $P(X = 10) = -0.1$, which violates one of the properties of a probability distribution, namely, that probabilities are nonnegative.

b. This is not a probability distribution because the sum of the probabilities for all values of X is more than 1, instead of being exactly 1.

c. This is a probability distribution because all probabilities are nonnegative and sum to unity.

3. Find the expected value, variance, and standard deviation of the number of heads obtained in tossing a coin three times (or three coins once).

SOLUTION:

Tossing of a coin three times results in the following probability distribution. Assume the coin is unbiased. Here X refers to number of heads.

Table 6.1

X	P(X)	XP(X)	(X − 1.5)	(X − 1.5)²	(X − 1.5)²P(X)
0	1/8 = 0.125	0	−1.5	2.25	0.281
1	3/8 = 0.375	0.375	−1.5	0.25	0.281
2	3/8 = 0.375	0.750	0.5	0.25	0.281
3	1/8 = 0.125	0.375	1.5	2.25	0.281
Total	1.00	1.5	0	5	0.75

From the above table, we see that the mean or the expected value $E(X) = \mu = \sum XP(X) = 1.5$; the variance $= \sigma^2 = E(X - \mu)^2 = \sum (X - 1.5)^2 P(X) = 0.75$; and the standard deviation $= \sigma = \sqrt{\sigma^2} = \sqrt{0.75} = 0.866$.

5. Some insurance agents are paid a basic salary plus a commission based on the number of customers. Assume that the annual salary is given by the relation Y ($) = 80,000 + 0.4(100X), where X is the number of customers serviced by the agent and each customer pays $100 to the company from which the agent gets 40%. Based on prior experience, the underwriting department of the insurance company estimated the probability distribution of X (shown below).

Number of Customers

X (in thousands)	1	2	3	4
P(X)	0.2	0.3	0.3	0.2

Find the expected annual salary of an insurance agent and its standard deviation.

SOLUTION:

The probability distribution of X, the number of customers, is

Table 6.3

X	P(X)	XP(X)	(X − 2.5)	(X − 2.5)²	(X − 2.5)²P(X)
1	0.2	0.2	−1.5	2.25	0.45
2	0.3	0.6	−0.5	0.25	0.075
3	0.3	0.9	0.5	0.25	0.075
4	0.2	0.8	1.5	2.25	0.45
Total		2.5			1.05

Therefore,

$$E(X) = 2.5 \text{ or } 2500, \text{ since X is in thousands.}$$

The expected annual salary of the insurance agent is

$$E(Y) = E(80,000) + 0.4(100E(X))$$
$$= 80,000 + 40(2500)$$
$$= 80,000 + 100,000 = \$180,000.$$

$$V(Y) = (0.4)^2 \times 100^2 V(X) = 0.16(10,000)(1.05)$$
$$= 1680.$$

Therefore, the standard deviation of the annual salary of the insurance agent is

$$\sqrt{1680} = 40.99.$$

EXCEL APPLICATIONS

7. Answer the following questions for the gross rent data given below.

   a. We have generated the following discrete probability distribution. Compute the mean, variance, and standard deviation for this distribution by using the Function Key of Excel.

   b. Using Excel, draw the probability mass function for this distribution. Use the X values on the X-axis and probability, P(X), on the Y-axis.

Table 6.4   Gross Rent

Gross Rent ($)	X($) (Class Mark)	Renter-Occupied Units (Frequency)	Probability P(X) (Relative Frequency)
Less than $200	100*	1,447,098	0.04
$200–299	250	1,537,171	0.05
$300–499	400	6,733,722	0.20
$500–749	625	11,905,100	0.35
$750–999	875	7,058,164	0.21
$1000–1499	1250	3,835,217	0.11
$1500 or more	1750*	1,369,610	0.04
Total		33,886,082	1.00

Note: The starred values of X, the class marks, are used due to open-ended classes.
Source: 2001 Supplementary Survey: U.S. Census Bureau, Table 4, Profile of Selected Housing Characteristics. www.census.gov/acs/www/products/profiles/single/2001/ss01/tabular/010/01000us4.htm

SOLUTION:

a.

Table 6.5

X ($)	P(X)	XP(X)	(X − 706.5)	(X − 706.5)^2	(X − 706.5^2)*P(X)
100	0.04	4	−606.5	367,842.25	14,713.69
250	0.05	12.5	−456.5	208,392.25	10,419.6125
400	0.2	80	−306.5	93,942.25	18,788.45
625	0.35	218.75	−81.5	6642.25	2324.7875
875	0.21	183.75	168.5	28,392.25	5962.3725
1250	0.11	137.5	543.5	295,392.25	32,493.1475
1750	0.04	70	1043.5	1088892.25	43,555.69
Total		706.5			128,257.75

The mean = $706.50
Variance = $128,257.75 in squared dollars
Standard deviation = $358.1309

b.

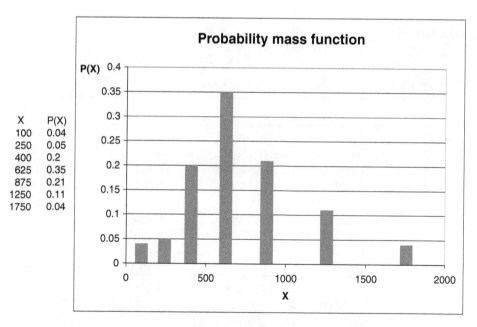

X	P(X)
100	0.04
250	0.05
400	0.2
625	0.35
875	0.21
1250	0.11
1750	0.04

9. Answer the following questions for the housing characteristics data given below.

   a. We have generated the following discrete probability distribution. Compute the mean, variance, and standard deviation for this distribution by using the Function Key and the long and the short formulas using Excel.

   b. Using Excel, draw the probability mass function for this distribution. Use the X values on the X-axis and probability, P(X), on the Y-axis.

Table 6.8 Housing Characteristics: Rooms

Rooms (X)	Number of Houses (Frequency)	Probability P(X) (Relative Frequency)
1	1,856,987	0.01576
2	4,400,070	0.03734
3	11,630,310	0.09871
4	20,958,754	0.17788
5	25,986,847	0.22056
6	21,886,935	0.18576
7	13,898,567	0.11796
8	8,876,069	0.07533
9 or more (let X = 10)	8,329,595	0.07070
Total	117,824,134	1.00

Source: 2001 Supplementary Survey: U. S. Census Bureau, Table 4, Profile of Selected Housing Characteristics. www.census.gov/acs/www/products/profiles/single/2001/ss01/tabular/010/01000us4.htm

SOLUTION:

a.

Table 6.9

X	P(X)	XP(X)	$P(X)^*(X - \mu)^2$	$X^2P(X)$
1	0.015761	0.0157607	0.31221	0.015761
2	0.037344	0.0746888	0.44469	0.149378
3	0.098709	0.2961272	0.59288	0.888382
4	0.177882	0.7115267	0.37440	2.846107
5	0.220556	1.1027812	0.04482	5.513906
6	0.185759	1.1145561	0.05603	6.687337
7	0.11796	0.8257219	0.28311	5.780053
8	0.075333	0.6026656	0.48955	4.821325
10	0.070695	0.7069515	1.46306	7.069515
Total	1	5.4507796	4.061	33.77176

Mean (μ) = 5.4507796.
Variance by long formula = 4.061.
Variance by short formula = $33.77176 - (5.4507796)^2 = 4.061$.
Standard deviation = $\sqrt{4.061} = 2.015$.

b.

X	P(X)
1	0.015761
2	0.037344
3	0.098709
4	0.177882
5	0.220556
6	0.185759
7	0.11796
8	0.075333
10	0.070695

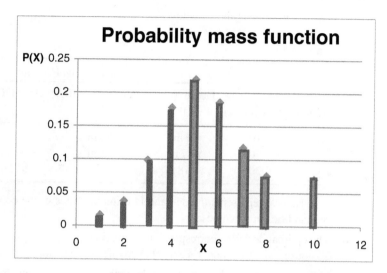

11. A contractor builds expensive homes, each costing $1.1 million, and lists them at a sale price of $1.5 million. His options are to build subdivisions of 10 homes at a time, 20 homes, or 30 homes depending on the demand for real estate. The demand for real estate housing depends on the state of the economy. The probabilities of different states of the economy are estimated as shown below.

State of the Economy

State of the Economy (X)	P(X)
Down	0.2
Same as now	0.3
Boom	0.5

Here X represents the demand for luxury homes.

a. Construct a payoff table.

b. Find the optimal decision based on expected payoff.

c. Construct an opportunity loss table.

d. What is your optimal decision based on expected opportunity loss?

SOLUTION:

a. The payoff table is shown below.

Table 6.12

Action	Demand (X)	State of the Economy			Exp. Payoff
		Down (0.2)	Same as Now (0.3)	Boom (0.5)	
		10	20	30	
A_1	10	$4 m	$4 m	$4 m	$4 m
A_2	20	$-7 m	$8 m	$8 m	$5 m
A_3	30	$-18 m	$-3 m	$12 m	$1.5 m

Here X represents the demand in different states of the economy, with their probabilities in parentheses.

b. The optimal decision, corresponding to maximum expected payoff, is to construct 20 houses, and the builder's expected payoff is $5 million.

c. Before constructing the opportunity loss table, we have to construct the maximum potential profit (corresponding to each possible state of the economy) table, which is as follows:

X	10	20	30
A(10)	4	8	12
A(20)	4	8	12
A(30)	4	8	12

Now, the opportunity loss (the difference between the maximum potential profit for each possible state of the economy and the net profit received for each chosen action) table is as follows:

A(10)	4 − 4	8 − 4	(12 − 4)
A(20)	4 − (−7)	8 − 8	(12 − 8)
A(30)	4 − (−18)	8 − (−3)	(12 − 12)

P(X)	0.2	0.3	0.5	Exp. Loss
A (10)	0	4	8	5.2
A (20)	11	0	4	4.2
A (30)	22	11	0	7.7

d. Based on the opportunity loss table, the optimal decision corresponding to minimum expected opportunity loss is A(20), for which minimum expected opportunity loss is $4.2 million.

Note that this is the same optimal decision based on the payoff table. Also note that none of the opportunity losses is less than zero.

13. The stock price of a company goes up or down by the close of the day every day. It is known from past experience that on the average the price goes up 20% of the time. Find the probability that the stock price goes up: (a) exactly 2 days, (b) at least 2 days, (c) at most 2 days, and (d) not once in the last four days, using the binomial formula.

SOLUTION:

Here we use the binomial probability distribution with parameters $n = 4$, and $p = 0.2$. Let X denote the number of days the stock price goes up. Then,

(a) $P(X = 2 : n = 4, p = 0.2) = {}_4C_2(0.2)^2(1 - 0.2)^{4-2}$

$$= \frac{4 \times 3 \times 2 \times 1}{2 \times 1 \times 2 \times 1}(0.04)(0.64) = 6(0.0256) = 0.1536 \text{ or } 15.36\%.$$

(b) $P(X \geq 2 : n = 4, p = 0.2) = {}_4C_2(0.2)^2(1 - 0.2)^{4-2} + {}_4C_3(0.2)^3(1 - 0.2)^{4-3}$

$$+ {}_4C_4(0.2)^4(1 - 0.2)^{4-4} = \frac{4 \times 3 \times 2 \times 1}{2 \times 1 \times 2 \times 1}(0.04)(0.64)$$

$$+ \frac{4 \times 3 \times 2 \times 1}{3 \times 2 \times 1 \times 1}(0.008)(0.8) + \frac{4 \times 3 \times 2 \times 1}{4 \times 3 \times 2 \times 1 \times 0!}(0.0016)(1)$$

$$= 0.1536 + 0.0256 + 0.0016 = 0.1808 \text{ or } 18.08\%.$$

(c) $P(X \leq 2 : n = 4, p = 0.2) = {}_4C_2(0.2)^2(1 - 0.2)^{4-2} + 4C_1(0.2)^1(1 - 0.2)^{4-1}$

$$+ 4C_0(0.2)^0(1 - 0.2)^{4-0} = \frac{4 \times 3 \times 2 \times 1}{2 \times 1 \times 2 \times 1}(0.04)(0.64)$$

$$+ \frac{4 \times 3 \times 2 \times 1}{1 \times 3 \times 2 \times 1}(0.2)(0.512) + \frac{4 \times 3 \times 2 \times 1}{0! \times 4 \times 3 \times 2 \times 1}(1)(0.4096)$$

$$= 0.1536 + 0.4096 + 0.4096 = 0.9728 \text{ or } 97.28\%.$$

(d) $P(X = 0 : n = 4, p = 0.2) = {}_4C_0(0.2)^0(1 - 0.2)^{4-0} = 0.4096$ or 40.96%. This probability is calculated in (c) above.

15. Based on historical information, it was found that approximately 10% of female employees with MBA degree and with job experience of at least 10 years have become the CEO's of major corporations. Consider a group of 15 female employees in U.S. major corporations with an MBA degree and having 10 years of job experience. Answer the following questions:

a. Explain how this example fits the binomial distribution via the three basic characteristics of the binomial distribution.

b. Find the parameters of the binomial distribution.

c. Find the probability that exactly 3 out of the 15 will become CEO's of major corporations.

d. Find the probability that at most three will become CEO's.

e. Find the probability that at least four will become CEO's.

f. Find the mean and standard deviation of the distribution and interpret your values.

(*Hint:* First show the binomial probability function for $n = 15$ and $p = 0.1$ and find the probability using the tables for binomial probabilities.)

SOLUTION:

a. This example fits the binomial distribution because there are two possible outcomes if you consider female employees with 10 years of job experience: (a) either she is a CEO (success) or not a CEO (failure); (b) the probability of such a female employee being a CEO (success) is given as 0.1, and the probability of not being a CEO (failure) is $1 - 0.1 = 0.9$; (c) the selection of a group of $n = 15$ female employees in major U.S. corporations with an MBA and having 10 years of experience is random.

b. The two parameters of the binomial distribution under consideration are $n = 15$ and $P = 0.1$, the probability of success.

c. $P(X = 3)$, where X is the number of female CEOs, using the binomial tables (for $n = 15$ and $r = 3$ in the column for 0.1) is 0.1285.

d. $P(X \le 3) = P(X = 0) + P(X = 1) + P(X = 2) + P(X = 3) = 0.2059 + 0.3432 + 0.2665 + 0.1285 = 0.9441$ using the same tables as above.

e. $P(X \ge 4) = 1 - P(X \le 3) = 1 - 0.9441$ (calculated above) $= 0.0559$ using the rule of complements in probability.

f. Mean $= \mu = n \times p = 15 \times 0.1 = 15$ and standard deviation $= \sigma = \sqrt{np(1 - p)}$ $= \sqrt{15 \times 0.1 \times (1 - 0.1)} = 1.162$.

Interpretation: In a group of 15 female employees with an MBA and having 10 years of job experience in a U.S major corporation, the average number of CEOs is 1.5 with a standard deviation of 1.162.

Index

Solutions Manual to Accompany Introduction to Quantitative Methods in Business: With Applications Using Microsoft® Office Excel®, First Edition. Bharat Kolluri, Michael J. Panik, and Rao N. Singamsetti.
© 2017 John Wiley & Sons, Inc. Published 2017 by John Wiley & Sons, Inc.

CPSIA information can be obtained
at www.ICGtesting.com
Printed in the USA
FSHW020836240521
81658FS